Multi-Agent Search Under Uncertainty

Multi-Agent Search Under Uncertainty

Reactive and Deep Q-Learning Methods

Barouch Matzliach
Tel Aviv University, Israel

Evgeny Kagan
Ariel University, Israel

Irad Ben-Gal
Tel Aviv University, Israel

WILEY

Registered Offices
John Wiley & Sons, Inc., 111 River Street, Hoboken, NJ 07030, USA
John Wiley & Sons Ltd, New Era House, 8 Oldlands Way, Bognor Regis, West Sussex, PO22 9NQ, UK

For details of our global editorial offices, customer services, and more information about Wiley products visit us at www.wiley.com.

The manufacturer's authorized representative according to the EU General Product Safety Regulation is Wiley-VCH GmbH, Boschstr. 12, 69469 Weinheim, Germany, e-mail: Product_Safety@wiley.com.

Wiley also publishes its books in a variety of electronic formats and by print-on-demand. Some content that appears in standard print versions of this book may not be available in other formats.

Library of Congress Cataloging-in-Publication Data applied for:

Harback ISBN: 9781394418459

Cover Design: Wiley
Cover Image: aerogondo/Getty Images

Set in 9.5/12.5pt STIXTwoText by Straive, Chennai, India

To the challenges that inspired us to solve problems, and to our loved ones – parents, wives, and children – who, each in their own way, are searching for their targets while standing by us as we search for ours.

Contents

Preface *ix*
About the Companion Website *xi*

1 Introduction *1*
1.1 The Problem of Search Under Uncertainty *1*
1.2 Book Objectives and Main Contributions *2*
1.3 Structure of the Book *3*

2 Background *5*
2.1 Probabilistic Search for Static and Moving Targets *6*
2.2 Search in Shadowed Space *10*
2.3 Search with False-Positive and False-Negative Errors *16*
2.4 Conclusions *19*

3 Problem Formulation and Basic Search Procedure *21*
3.1 Basic Assumptions *21*
3.2 Sensor's Fusion and Dynamic Probability Map *24*
3.3 Search Policy and Sensing *26*
3.3.1 Search Policy *26*
3.3.2 Search Time and Sensors' Sensitivity *27*
3.3.3 Numerical Simulations *28*
3.4 Conclusions *30*

4 Reactive Detection Algorithms *31*
4.1 Agents' Policies and Decision-making *31*
4.2 Policies Control and Brute-Force Learning *34*
4.3 Numerical Simulations *35*
4.3.1 Detection by a Single Agent *36*
4.3.2 Detection by Multiple Agents *38*
4.4 Conclusions *42*

5 Search with Learning *43*
5.1 Reinforcement Learning *43*
5.2 Search with Supervised and Unsupervised Learning *47*

5.3 Two-Actor Search *50*
5.3.1 Pursuit-Evasion Game *51*
5.3.2 Actor-Critic Approach *52*
5.4 Conclusions *53*

6 Search with Deep Q-Learning: Single Agent *55*
6.1 Problem Formulation and Notation *55*
6.2 Decision-making Policy and Deep Q-Learning Solution *58*
6.2.1 Agent's Actions and Decision-making *58*
6.2.2 Dynamic Programming and Target Neural Networks *59*
6.2.3 Model-Free and Model-Based Learning *60*
6.2.4 Choice of the Actions at the Learning Stage *63*
6.2.5 The Q-max Algorithm *63*
6.2.6 The SPL Algorithm *67*
6.3 Numerical Simulations *67*
6.3.1 Network Training in the Q-max Algorithm *68*
6.3.2 Detection by the Q-max and SPL Algorithms *69*
6.3.3 Comparison Between Q-max and SPL Algorithms and One-Step Heuristics *69*
6.3.4 Comparison Between SPL Algorithm and Optimal Solution *73*
6.3.5 Run Times and Mean-Squared Errors for Different Sizes of Data Sets *75*
6.4 Conclusions *76*

7 Search with Deep Q-learning: Multiple Agents *77*
7.1 Problem Formulation and Notation *77*
7.2 Cooperative Detection Using Voronoi Regions and Deep Q-Learning *78*
7.2.1 Agents' Actions and Decision-making *79*
7.2.2 Reactive Decision-making in Voronoi Regions: A Distributed *EIG* Algorithm *79*
7.2.3 Collective Deep Q-Learning Approach *81*
7.3 Numerical Simulations *88*
7.3.1 Detection of Static Targets *89*
7.3.2 Detection of Moving Targets *92*
7.3.3 Learning Errors and Run Time of the Collective Q-max Algorithm *94*
7.4 Conclusions *95*

8 Conclusions *97*

References *99*
Index *103*

Preface

In 1977, Lawrence Stone published the paper entitled *Search Theory: The Mathematical Theory for Finding Lost Objects*, and this title essentially defines the search theory. This theory includes a wide range of methods and algorithms for navigating the search agents looking for hidden targets. In different applications, the algorithms deal with static and moving targets, with the targets escaping from the searchers and with the searchers acting in different values of uncertainty.

In search theory, an enormous number of papers and several books were published, and different algorithms were developed, including algorithms that optimally solve specific problems of search and heuristic algorithms that treat real-world problems. Nevertheless, up to now, the solution of each practical problem requires specific analysis, correct selection, and adjustment of the algorithms, which is rather problematic.

Once while thinking about the problem, we asked the AI assistant:

- What is search in the search theory?

And it replied:

- Search is not just movement in space, but the art of allocating attention.

The phrase "allocating attention" in contrast to the usual "achieving maximal expected reward or minimal expected payoff" seemed to be interesting and inspired us to consider some algorithms that follow this approach in one volume. The algorithms were developed by our group in Tel-Aviv and Ariel Universities during the last decade and proved their efficiency in different applications.

In this book, we present heuristic algorithms for navigation of the search agents using immediate information about the targets' location probabilities and heuristic algorithms with deep Q-learning. These algorithms follow different ideas but have a common property: they do not implement direct optimization but distribute the searchers' attention over the domain.

In certain sense such an approach is a return to the origins of the Koopman search theory, which addressed distribution of the search efforts in the framework of path-planning and search by mobile search agents.

The book is self-contained and can be read independently of the other sources in the theory of search. Along with that, it cannot be considered as a textbook in the field of search or an overview of the theory. Our goal is to provide additional material that will bridge the gaps in search theory and will be useful for practical applications.

We hope that together with the books by Stone and Washburn and their colleagues, this book will be useful for STEM students and researchers and will contribute to the progress in search theory and its applications.

This book is the result of the long and intensive work of many people – from our teachers and colleagues, who showed us, whenever possible, how to better understand the world, to our students, who helped us identify the gaps in our understanding.

We thank Prof. Oded Maimon, Prof. Lev Levitin, Prof. Michael Caramanis, Prof. Yigal Gerchak, Prof. Evgeny Khmelnitsky, Prof. Yossi Bukhchin, Mr. Ofer Tzimchi and other friends in Shaldag, Prof. Victor Kureychik, Prof. Askold Melikhov, Prof. Yuri Kuznetzcov, Prof. Vladimir Rijov, Prof. Shraga Shuval, and Prof. Nir Shvalb, who have influenced, and continue to influence, our research.

We thank Boaz Herskovitz, Gal Goren, Moshe Israel, and Yuval Ziv, who participated in different stages of this research and whose contributions were crucial to the success of the methods presented in this book.

We thank the Wiley team for their attention and kind support. This is the third book we have published with Wiley, and once again we are pleased to work with these excellent professionals.

Above all, nothing could happen without the loving support of our families, who stand by us at every moment and under all circumstances. To our parents, children, and wives, Revital, Galia, and Hila – thank you!

Tel Aviv, Israel
March 2026

Barouch Matzliach
Evgeny Kagan
Irad Ben-Gal

About the Companion Website

This book is accompanied by a companion website.

www.wiley.com/go/matzliach/multiagentsearch1

To access the website materials, click on the link under Related Resources on the book's page on wiley.com. You can also find this by searching www.wiley.com for the ISBN 9781394418459.

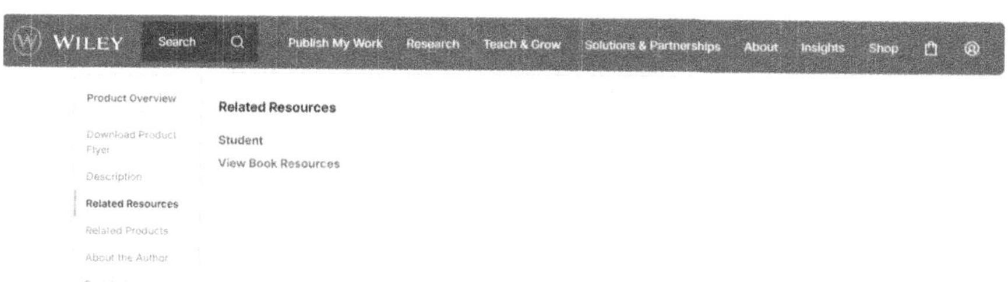

The URL above will be redirected to the author's site below.

https://github.com/Barouchm/SearchProblem

1 Introduction

The first mathematical formulations of the problem of search appear at the end of the 17th century in the works by Huygens and Leibnitz [43] who required to find a trajectory of the searcher who pursues the target and tries to chase it. In the 20th century, this problem was considered both in its original formulation [61] and as a search game [27] in which the target also sees the searcher and tries to escape.

The other formulation of the problem was suggested in 1942 by Koopman [39] who considered distribution of search efforts over the domain and suggested his famous random search formula, which originated probabilistic search methods. In this formulation the goal is to detect the target or, in other words, to recognize location of the hidden target. Later Stone [55] suggested the theory of optimal search, which together with the Washburn pursuit theory [61] forms a basis for the development of search algorithms.

In this book, we consider the problem of search for static and moving targets by a single searcher and by a group of searchers. We assume that the searchers have imperfect information about locations of the targets and need to handle this uncertainty. The goal is to find a search policy such that the targets are detected in minimal time.

1.1 The Problem of Search Under Uncertainty

Modern studies of the methods of search for hidden objects follow two main directions. The first was originated in 1942 by the Koopman and the US Navy's Antisubmarine Warfare Operations Research Group with the aim of detecting German submarines in the Atlantic [39, 40]. As was later formulated by Frost and Stone [15, p. 13], the main question of these studies is

> how to most effectively employ limited resources when trying to find an object whose location is not precisely known. The goal is to deploy search assets to maximize the probability of locating the search object with the resources available. Sometimes this goal is stated in terms of minimizing the time to find the search object.

In the original scenario considered by Koopman, the searcher starts with a certain distribution of probabilities of the target's location in the domain and is required to distribute the search efforts over a domain, given these probabilities. It was assumed that the target is static and that the search efforts are infinitely divisible. After distribution of the search efforts, the points of the domain are observed, and the target is either detected or not. The goal is to find such a distribution of the search efforts that the probability of detecting the target is maximal.

To solve this problem, Koopman suggested his famous random search formula [39]

$$\varphi(x,t) = 1 - \exp(-\kappa(x,t)),$$

which defines probability $\varphi(x, t)$ of detecting the target in the point x at time t with respect to the search effort $\kappa(x, t)$ applied to the point x at time t and given that the target is located in the point x.

The Koopman random search formula states that the probability of detecting the target at a certain point is as great as the applied search effort, and perfect detection is resulted by application of infinite search efforts.

Starting with the Koopman research, in 1975 Stone [55] published the theory of optimal search and suggested an algorithm for creating optimal distribution of search efforts over the search domain. This algorithm considers the distribution of the probabilities of target's locations and iteratively defines the quantity of the search efforts that should be applied to each point of the domain. The Stone algorithm implements optimization using Lagrange multipliers and starting from the first point iteratively assigns the value of search efforts to the points of the domain proportionally to the logarithm of the location probabilities.

In 1980, Washburn [59, 60] developed the algorithm that defines optimal distribution of search efforts in the search for a moving target. This algorithm considers the probabilities of detecting and non-detecting the target and follows, respectively, forward and backward recursive equations for calculating these probabilities. Distribution of search efforts is defined by inverse Koopman formula such that it maximizes the probability of catching the target. As a result, given the probabilities of the target's movements, the algorithm creates steady-state distribution of the search efforts.

The second direction in the studies of search methods considers mobile searchers moving over the domain and seeking hidden static or moving targets. In contrast to the search games, in these studies it is assumed that the targets are not aware of the searchers' actions. Detection of the targets does not require catching the target and results from observation of the parts of the domain. Detection probability depends on the applied detection function that represents the used sensors.

For arbitrary function of detection probability in 1969, Ross [51] suggested a model that follows the direction of statistical decision processes and uses dynamic programming optimization. Later such models were formulated in terms of Markov Decision Processes [64] that allowed to consider imperfect detections with the errors of the first and the second type and to study the search for static and moving targets.

Solution for the simplest scenario of search for moving target was published in 1970 by Pollock [47], who considered a search for a target moving between two boxes and found an optimal policy for the searcher moves such that the target will be caught in minimal time. In later studies, the Pollock approach was extended for the domain with arbitrary number of points (for brief overview and appropriate references, see e.g. [35]).

In recent books, Stone et al. [56] and Washburn and Kress [63] considered both the indicated algorithms of probabilistic search and their extensions and suggested algorithms of search for moving targets [57], which combine the Koopman approach and the search by mobile agents. Along with that, a general solution to the search problem was not suggested yet, and the statement by Ross [52] that the searching for a moving target is an open problem remains valid.

1.2 Book Objectives and Main Contributions

The book addresses the search for static and mobile targets by the group of searchers, which are considered as autonomous agents – mobile robots equipped with certain sensors and governed by onboard controllers. It is assumed that in the group the agents can communicate with each other and share information about the targets detected.

In this book, we consider two types of algorithms:

- *reactive algorithms*, which govern the agents to plan their next actions following the information perceived by the sensors,
- *deep Q-learning algorithms*, which govern the agents to learn the patterns of the targets' motion and to plan their actions with respect to the learned patterns.

It is assumed that the sensed information is not always correct and that detection allows both false-negative and false-positive errors, which means that there is a nonzero probability that the agent does not recognize the existing target and that there is a nonzero probability that the agent recognizes nonexistent target.

The goal of the book is to present recently developed procedures for navigation of mobile agents searching for multiple targets in the presence of detection errors of the first and the second types. These procedures were developed with respect to different detection tasks and scenarios and are implemented in the form of constructive algorithms for detecting multiple static and mobile targets by the group of mobile agents acting under uncertainty.

The presented *reactive online algorithms* implement the strategies based on the "center of gravity" of the probability map and on the "center of view" of the used sensors and plan one-step or multi-steps motion using expected information gain as a decision-making criterion. The most effective reactive algorithm is an algorithm that implements dynamic Voronoi partitioning of the search domain and maximizes a one-step expected information gain in each Voronoi region.

In the presented *algorithms with deep Q-learning* abilities, it is assumed that the search agents are equipped with the strong enough computers processing prediction and target neural networks, which allow learning of the targets' activity. In the single-agent search, both networks are updated by the agent using the information obtained by its sensors. In the multi-agent search, the agents are allowed to communicate such that the next-in-line neighboring agent updates its prediction network with respect to the target network of the previous-in-line agent. Training of the networks can be conducted offline with an online correction or online starting with certain initial probability maps, which are corrected during the search.

The algorithms of both types are applicable in the search for multiple static and moving targets in the presence of the sensing errors of the first and the second types.

1.3 Structure of the Book

The book includes eight chapters, and the rest of the book is organized as follows. First, we form a background for further material and present three algorithms of search for static and moving targets under uncertainty, which illustrate the considered problem. Then, we formulate the problem and introduce the required terminology and basic concepts. The next chapter addresses reactive search procedures, and the final part deals with the algorithms with deep Q-learning.

More precisely, the contents of the chapters are the following.

Chapter 2, *Background*, briefly describes the considered search and detection problem and includes brief literature review. In addition, it includes three reactive algorithms of search: the first is a simple algorithm with information criterion, which resolves the uncertainty using diffusion process; the second is an algorithm of search for static targets with detection errors of the first and the second types; and the third is the algorithm of search in shadowed space.

Chapter 3, *Problem Formulation and Basic Search Procedure*, includes formalization of the considered problem of search and detection in the gridded domain, which is used in all scenarios throughout the book. Also, the chapter presents exact procedures for updating the probability maps and sensor fusion and specifies relations between probability maps of different levels, dynamics of sensors sensitivity, and communication between the agents. Finally, it presents the basic search procedure in the presence of false-positive and false-negative sensing errors.

Chapter 4, *Reactive Detection Algorithms*, presents the extensions of the formulated basic procedure and considers reactive algorithms of search. In these algorithms, the search agents are governed toward either the "center of gravity" of the probability map or the "center of view" of the sensors with one-step or multi-steps motion planning. Decision-making in both cases is based on the expected information gain. These algorithms are applied for the search by a single agent and by the groups of agents.

Chapter 5, *Search with Learning*, starts the consideration of search algorithms with the ability to learn the targets' motion patterns. The chapter presents a brief overview of supervised and unsupervised learning in the framework of search by mobile agents. This preliminary information is used in the next two chapters.

Chapter 6, *Search with Deep Q-learning: Single Agent*, presents the problem of search in the form of dynamic programming procedure. Based on this procedure, a deep reinforcement learning algorithm of search by a single agent is formulated. This algorithm implements deep Q-learning, which is used for maximizing cumulative gain of information about targets' locations and for minimizing the length of the searcher's trajectory. The algorithm terminates while a predefined detection probability is reached.

Chapter 7, *Search with Deep Q-learning: Multiple Agents*, presents an extension of the algorithm considered in the previous chapter to the search by multiple agents. To obtain such an extension, at first, a reactive algorithm based on the dynamic Voronoi partitioning of the search space is formulated. This algorithm plans the agents' paths such that they maximize the expected one-step look-ahead information in the agents' Voronoi regions. Then, the algorithm of search with deep Q-learning is formulated. In this algorithm, it is assumed that each search agent controls its prediction and training networks and that each agent updates its prediction network using the information stored in the target network of its previous-in-line neighboring agent.

Chapter 8, *Conclusions*, summarizes the discourse and presents further directions of the research.

All chapters that consider algorithms or search include numerical simulations and examples.

The book is self-contained and includes ready-to-use algorithms. Along with that, we consider it as an element of the array of publications, which fills a certain gap in the search theory but does not present complete search theory or even introduction to it.

To obtain additional information on the search theory, we recommend the report [7] and the report and the books by Stone and his colleagues [15, 55, 56]; the report [62] and the books [61, 63] by Washburn and his colleague; the book by Stone, Royset, and Washburn [57]; and our books [35, 36]. These books also include descriptions and references to the studies of the other approaches to search problems: bio-mimetic search, search in continuous space and time, and search based on path-planning algorithms.

The Python code of the main algorithms considered in the book can be downloaded from the page [20].

2

Background

The book follows conventional methodology of the theory of search and detection [41, 55, 61] and probabilistic and informational methods of tracking targets under uncertainty [35, 56, 57].

In these classical formulations, search problems address the scenarios in which the searcher pursues the target, and solution of the problem is the shortest trajectory of the searcher. Detection problems, in contrast, focus on recognition of the target's location without necessarily reaching it. Often, detection is considered the separate task that precedes chasing of the targets by the search agent. For an overview of the problems and results, see e.g. [2, 15].

As indicated above, in 1975, Stone published the main concepts and ideas of the search theory [55] and initiated the use of optimization techniques for distributing search efforts. In 1989, Washburn presented formal techniques for tracking targets by mobile agents [61]. Later, the methods of search and screening and the methods of target detection and tracking were unified into a general probabilistic framework [56, 57].

Further developments in the theory of search addressed the search for several targets by a group of mobile search agents. The agents start at certain locations, move over the domain, and explore it up to the detection of all the targets. Along with obvious advantages, the use of such groups requires complex techniques of preserving group activity, collision avoidance, and communication. To overcome such a challenge, cooperative search methods implemented different heuristic approaches, see, e.g. [46, 50, 54], and learning techniques [9, 30, 44].

Standard taxonomies of this problem are based on the targets' and the searchers' abilities, such as their mobility, knowledge about the activities of the other party, and cooperation level [35, 37, 50]. Also, many search algorithms are classified with respect to the optimization principles that govern the agents' motion in the group. Often, search algorithms are implemented by using different heuristics, mostly informational ones, or by mimicking animal foraging [36, 37, 54].

In addition, the studies of collective search include considerations of communication and collective decision-making under uncertainties [2, 15], see also [37] and references therein. In practice, communication between the agents and information processing can be organized at different levels: from peer-to-peer networks up to transmitting information by a central station that receives. In the first case, each agent obtains information from its neighbors and makes decisions based on such local information. In the second case, a central station defines the agents' motion based on global information. In practice, the algorithms of control of the agents group use both approaches.

In this book, we consider a detection process in which a group of search agents moves over the domain and searches for a single or several static or moving targets. Following the basic Koopman formulation, we consider a probabilistic search scheme in which the search agents have knowledge only about the probabilities of the targets' location. Also, we assume that the Koopman exponential random search formula, which defines detection probabilities, is applicable.

Formally, such a problem can be considered as a problem of navigation in a dynamic environment with probabilistic navigation function [21–23]. However, since in contrast to the goal of navigation function used to govern the robots to avoid obstacles and the goal of search algorithms is to move the robots toward the targets, the target location probabilities can be considered as the complements of the probability navigation functions.

Decisions about the next step of each agent are based on the expectations about targets' locations, detection results, and detection errors and the steps of the other agents. The method that counts these expectations in real time uses the method of the Voronoi diagrams and estimates the further observations in the Voronoi areas of the agents [10]. We use such techniques in the reactive algorithm with distributed decision-making.

Additional problem that appears in the search by a group of mobile agents is coordination of simultaneous and sequential actions [46, 58], which is solved either by specification of the type of activity, which requires strong synchronization of the agents' behavior, or by alternating the actions' order, which requires additional rules for choosing active agents.

In this book, we continue a line of practical algorithms, and in contrast to the known methods of collective search and detection, consider a more realistic situation of search with false-positive and false-negative detection errors [5, 34].

The other practical assumption is the existence of a variety of sensors. The implemented sensor-fusion scheme follows a general Bayesian scheme [56] with varying sensitivity of the sensors.

Similarly to most algorithms of search in a gridded domain, the presented algorithms follow the occupancy grid approach [12, 13] by which the probability map of the targets' candidate locations is created simultaneously with the detection process and the motion of the search agents [12, 13].

To illustrate the considered problem and introduce the main terms, we consider three simple algorithms of search. The first acts on the array of the targets' location probabilities with observation of the complete domain and without detection errors, the second acts in the shadowed domain without detection errors, and the third defines the search with the detection errors.

2.1 Probabilistic Search for Static and Moving Targets

Let us start with a simple reactive algorithm of search for static and moving targets [33, 34].

Denote by $C = \{c_1, c_2, \dots, c_n\}$ a finite set of cells, which represents a rectangle geographical domain of the size (n_1, n_2), $n = n_1 \cdot n_2$. A target can be located in the cells $c_i \in C$, $i = 1, 2, \dots, n$, at any time $t = 0, 1, 2, \dots$. We index the cells in linear order such that for the Cartesian coordinates (j_1, j_2) of the cell c_i, its index is given by $i = ((j_1 - 1)n_1 + j_2)$, where $j_1 = 1, \dots, n_1$ and $j_2 = 1, \dots, n_2$.

The target location probabilities at the time t, $t = 0, 1, 2, \dots$, are denoted by

$$p_i(t) = \Pr\{c(t) = c_i\}, \quad i = 1, 2, \dots, n,$$

where $c(t)$ denotes the target's location at time t. We assume that $\sum_{i=1}^{n} p_i(t) = 1$, which means that at any time t the target is located in some cell in C with certainty.

The target's movement is governed by a discrete Markov process with a transition probabilities matrix

$$\rho = [\rho_{ij}]_{n \times n}, \quad \rho_{ij} = \Pr\{c(t+1) = c_j | c(t) = c_i\},$$

where $\sum_{j=1}^{n} \rho_{ij} = 1$ for each $i = 1, 2, \dots, n$. If matrix ρ is a unit matrix, then the target is static, otherwise we say that it is moving.

The search agent moves over the set C and at each time t, $t = 0, 1, 2, \dots$, observes a neighboring area $O(t) \subset C$ of certain radius $r > 0$, which is defined by the used sensors and the metric of the considered domain.

The result of observation is specified by detection probability function φ, which specifies the probability $\varphi(O(t))$ of detecting the target in the observed area $O(t)$. In contrast to the Koopman search and screening procedure, in this algorithm we assume that the detection probability is defined by indicator function such that

$$\varphi(O(t)) = \mathbf{1}(c(t), O(t)) = \begin{cases} 1 & \text{if } c(t) \in O(t), \\ 0 & \text{otherwise.} \end{cases}$$

In other words, if the search agent observes an area $O(t)$ and the target is in some cell $c(t)$ in this area, then the agent with certainty decides that the target is somewhere in the area $O(t)$.

The search is conducted as follows [32]. The search agent deals with the expected location probabilities $\tilde{p}_i(t)$ that at initial time $t = 0$ are equal to the initial location probabilities $\tilde{p}_i(0) = p_i(0)$, $i = 1, 2, \dots, n$.

At each time t, the agent chooses an area $O(t) \subset C$ and observes it. If the probability $\varphi(O(t)) = 1$, then the search terminates. Otherwise, the agent updates expected location probabilities $\tilde{p}_i(t)$ over the set C. The updated probabilities are called observed probabilities and are denoted by $\overline{p}_i(t)$.

To the observed probabilities $\overline{p}_i(t)$, the agent applies a known or estimated rule of the target movement and obtains the expected location probabilities $\tilde{p}_i(t+1)$, $i = 1, 2, \dots, n$, for the next time $t + 1$. For example, if the target moves according to a Markov process with a transition probabilities matrix ρ, then the expected location probabilities for the next time $t + 1$ are defined by

$$\tilde{p}_i(t+1) = \sum_{j=1}^{n} \rho_{ij} \overline{p}_i(t), i = 1, 2, \dots, n.$$

Given the expected probabilities $\tilde{p}_i(t+1)$, the agent chooses the next area $O(t+1)$ and the process continues.

The goal is to find a policy for choosing the areas $O(t)$, $t = 1, 2, \dots$, such that the search procedure terminates in a minimal average (over possible target moves) number of steps.

In the algorithm, we implement a Manhattan metric and assume that at each time t, both the searcher and the target can apply one of the following movements:

$$\mathbb{a}(t) \in \mathbb{A} = \{\uparrow, \rightarrow, \downarrow, \leftarrow, \odot\}$$

where arrows define the directions of the steps and $\odot$ means "to stay in the current position"; later we will also consider the diagonal movements.

The target's choice of movements is conducted according to its transition probabilities matrix ρ, while the agent's choice follows the decision that is obtained according to the values $\varphi(O(t))$ and the expected probabilities $\tilde{p}_i(t+1)$.

Formally, the algorithm is outlined as follows.

Algorithm 2.1 Search with errorless detection

Input: Set $C = \{c_1, c_2, \dots, c_n\}$, probabilities $p_i(0)$, $i = 1, 2, \dots, n$, matrix $\rho = [\rho_{ij}]_{n \times n}$, and the set $\mathbb{A}$ of possible movements.

Output: Area $O(t) \subset C$ such that $\varphi(O(t)) = 1$.

---

1. Set $t = 0$.
2. Set $\tilde{p}_i(t) = p_i(0)$, $i = 1, 2, \dots, n$.

3. Choose $O(t)$ arbitrary.
4. While $\varphi(O(t)) \neq 1$, do

   Calculate probabilities $\overline{p}_i(t)$, $i = 1, 2, \ldots, n$:

5. If $c(t) \in O(t)$ then for all cells $c_j \in O(t)$ set $\tilde{p}_j(t) = 0$.
6. For each $i = 1, 2, \ldots, n$ set $\overline{p}_i(t) = \tilde{p}_i(t) / \sum_{j=1}^{n} \tilde{p}_j(t)$.

   Calculate probabilities $\tilde{p}_i(t+1)$, $i = 1, 2, \ldots, n$:

7. For each $i = 1, 2, \ldots, n$ set $\tilde{p}_i(t+1) = \sum_{j=1}^{n} \rho_{ij} \overline{p}_i(t)$.

   Calculate weights w_k for the movements $\mathbb{a}_k \in \mathbb{A}$:

8. For each $\mathbb{a}_k$ set $w_k = \sum_{c_j \in \tilde{O}_k} \tilde{p}_j(t+1)$, where $\tilde{O}_k$ is an area that will be observed if the agent make a step $\mathbb{a}_k$.

   Estimate global probabilities if need:

9. If $w_k = 0$ for each k then
10. Set $\tilde{\mathbb{p}}$ = *estimate probabilities*$(\tilde{p}(t+1), \rho)$,
11. Else
12. Set $\tilde{\mathbb{p}} = \tilde{p}(t+1)$,
13. End if.

   Apply local decision-making:

14. Set $O(t+1)$ = next observed area$(\tilde{\mathbb{p}})$.

   Move to the area $O(t+1)$.

15. Set $t = t+1$.
16. End while.
17. Return: the target is observed in the area $O(t)$.

The algorithm includes two key functions: *estimate probabilities*( ) function which defines expected probabilities over the domain and *next observed area*( ) function which implements decision making about the next observed area $O(t+1)$.

In the function next observed area( ) we implement local decision-making which is based on discrete version of maximum gradient over expected probabilities $\tilde{\mathbb{p}}_i$ from the current location of the search agent to possible observed areas. In other words, the agent chooses the next feasible observed area according to the defined movements $\mathbb{A}$ for which the observation will maximally influence to the probabilities $\tilde{\mathbb{p}}_i$, $i = 1, 2, \ldots, n$.

The distances between the probability mass functions $\tilde{\mathbb{p}}_k$, where k corresponds to the indices of the movements $\mathbb{a}_k \in \mathbb{A}$, and the probability mass function $\tilde{\mathbb{p}}$ are calculated as follows:

$$\mathrm{dist}(\tilde{\mathbb{p}}_k, \tilde{\mathbb{p}}) = \sum_{i=1}^{n} |\tilde{\mathbb{p}}_{ki} - \tilde{\mathbb{p}}_i|.$$

Given the set of possible movements $\mathbb{A}$, the function *next observed area*( ) is outlined as follows.

Algorithm 2.2 Function *next observed area*

Input: Array of expected probabilities $\tilde{\mathbb{p}} = (\tilde{\mathbb{p}}_1, \tilde{\mathbb{p}}_2, \ldots, \tilde{\mathbb{p}}_n)$, set of movements $\mathbb{A}$.
Output: Area $O \subset C$.

---

1. For all movements $\mathbb{a}_k \in \mathbb{A}$ do:
2. Determine the area O_k according to the movement $\mathbb{a}_k$.
3. Set $\mathbb{w}_k = \sum_{c_j \in O_k} \tilde{\mathbb{p}}_j$.
4. For each $c_i \in C$, if $c_i \in O_k$, then set $\tilde{\mathbb{p}}_{ki}^{1} = 1$.

5. For each $c_i \in C$, set $\overline{\wp}^1_{ki} = \tilde{\wp}^1_{ki} / \sum_{i=1}^{n} \tilde{\wp}^1_{ki}$.
6. For each $c_i \in C$, if $c_i \notin O_k$, then set $\tilde{\wp}^0_{ki} = 0$.
7. For each $c_i \in C$, set $\overline{\wp}^0_{ki} = \tilde{\wp}^0_{ki} / \sum_{i=1}^{n} \tilde{\wp}^0_{ki}$.
8. Set $d(O_k) = w_k dist\left(\overline{\wp}^1_{ki}, \tilde{\wp}\right) + (1 - w_k) dist\left(\overline{\wp}^0_{ki}, \tilde{\wp}\right)$.
9. End for.
10. Set $O = \text{argmax}\{d(O_k)$ over all $O_k\}$; ties are broken randomly.
11. Return O.

Decision-making function *next observed area*( ) acts on the probabilities $\tilde{\wp}_i$, $i = 1, 2, \ldots, n$. If these probabilities allow to treat the uncertainty and choose the next observed area, then these probabilities are equal to the probabilities $\tilde{p}_i(t+1)$, and if for all possible movements $\mathbb{a}_k \in \mathbb{A}$, the probabilities $\tilde{p}_i(t+1)$ are equal to zero, then global estimation of the probabilities is applied.

The probabilities estimation is conducted by the function *estimate probabilities*( ), which starts from the probabilities $\tilde{p}_i(t+1)$ and results in the probabilities $\tilde{\wp}_i$, $i = 1, 2, \ldots, n$. Probabilities $\tilde{\wp}_i$ are calculated using the expected target's transition probabilities matrix $\tilde{\rho}$ as the target location probabilities after a certain number m of steps.

For moving targets with known transition probabilities matrix ρ, the expected transition probabilities matrix $\tilde{\rho}$ is equivalent to ρ, and for static targets the matrix $\tilde{\rho}$ defines a process of simple random walk.

The number m of steps is defined using the estimated target's location and as an entropy measure [8], which specifies the lack of information available to the searcher. Since the search agent observes the areas O that include equal number of cells, the entropy is calculated using the size s of the areas. Then, the required or missing amount of information is defined as follows:

$$I(\tilde{p}(t+1)) = \log_s(n) - \sum_{i=1}^{n} \tilde{p}_i(t+1) \log_s(\tilde{p}_i(t+1)).$$

Note that the logarithms are base the size s of the observed areas O which follows the assumption that the areas O have the same size and that the observations are error free.

Given the current location $y \in C$ of the search agent, the probabilities $\tilde{p}_i(t+1)$, $i = 1, 2, \ldots, n$, and the target's transition probabilities matrix ρ, the function *estimate probabilities*( ) is outlined as follows.

Algorithm 2.3 Function *estimate probabilities*

Input: Array of expected probabilities $\tilde{p}(t+1) = (\tilde{p}_1(t+1), \tilde{p}_2(t+1), \ldots, \tilde{p}_n(t+1))$, expected transition probabilities matrix $\tilde{\rho} = [\tilde{\rho}_{ij}]_{n \times n}$.

Output: Expected probabilities $\tilde{\wp} = (\tilde{\wp}_1, \tilde{\wp}_2, \ldots, \tilde{\wp}_n)$.

---

1. Set $\tilde{c} =$ *estimated location*$(\tilde{p}(t+1))$: expected target's location.
2. Set $d = \text{dist}(\tilde{c}, y)$: distance between expected target's location and the current searcher's location.
3. Set $m = \text{ceil}(I(\tilde{p}^{t+1})) \cdot \text{round}(d)$: number of expected steps.
4. For each $c_i \in C$, set $\tilde{\wp}_i = \tilde{p}_i(t+1)$.
5. Do m times:
6. For each $c_i \in C$, set $\wp_i = \tilde{\wp}_i$.
7. For each $c_i \in C$, set $\tilde{\wp}_i = \sum_{j=1}^{n} \tilde{\rho}_{ij} \wp_i$.
8. End do.
9. Return $\tilde{\wp}$.

In this algorithm, round(b) stands for the closest integer number of the real number b, and ceil(b) stands for the closest integer number that is greater than b. For two cells $c = (j_1, j_2)$ and $c' = (j'_1, j'_2)$, function dist(c, c') calculates the distance between the cells. In the considered algorithm of search in discrete space, the distance is defined by the Manhattan metric

$$\text{dist}(c, c') = |j_1 - j'_1| + |j_2 - j'_2| .$$

Finally, function *estimated location*( ) returns the center of distribution $\tilde{p}(t+1)$, which is calculated as a "center of gravity" for the probabilities $\tilde{p}_i(t+1)$, $i = 1, 2, \ldots, n$. For the considered rectangle domain of the size (n_1, n_2), $n = n_1 \cdot n_2$, coordinates $(\tilde{j}_1, \tilde{j}_2)$ of the expected target's location $\tilde{c}$ are calculated as follows.

Algorithm 2.4 Function *estimated location*

Input: Array of expected probabilities $\tilde{p}(t+1) = (\tilde{p}_1(t+1), \tilde{p}_2(t+1), \ldots, \tilde{p}_n(t+1))$, coordinates $j_1 = 1, \ldots, n_1, j_2 = 1, \ldots, n_2$.

Output: The cell $\tilde{c} = (\tilde{j}_1, \tilde{j}_2)$.

---

1. Set $\tilde{j}_1 = 0$ and $\tilde{j}_2 = 0$.
2. For $j_1 = 1, \ldots, n_1$ do:
3.   For $j_2 = 1, \ldots, n_2$ do:
4.   Set $\tilde{j}_1 = \tilde{j}_1 + j_1 \cdot \tilde{p}_i(t+1)$, $i = ((j_1 - 1)n_1 + j_2)$.
5.   End for.
6. End for.
7. For $j_2 = 1, \ldots, n_2$ do:
8.   For $j_1 = 1, \ldots, n_1$ do:
9.   Set $\tilde{j}_2 = \tilde{j}_2 + j_2 \cdot \tilde{p}_i(t+1)$, $i = ((j_1 - 1)n_1 + j_2)$.
10.   End for.
11. End for.
12. Set $\tilde{j}_1 = round(\tilde{j}_1)$ and $\tilde{j}_2 = round(\tilde{j}_2)$.
13. Return $\tilde{c} = (\tilde{j}_1, \tilde{j}_2)$.

The snapshots of the algorithm's activity are shown in Figure 2.1.

In Figure 2.1a, the search agent arrives at the cell $c = (30, 8)$ such that the probabilities around this cell are equal to zero. Hence, it applies Algorithm 2.3 for estimating the probabilities. The probabilities diffuse over the domain; result of the probabilities' estimation is shown in Figure 2.1b. Then the agent decides about the next step following the estimated probabilities.

Algorithm 2.1 can be easily extended to the search by several searcher agents for several static or moving targets and even to the search with detection errors of the first type, and in all these scenarios it demonstrated rather effective search [33–35].

The ideas underlying Algorithm 2.1 formed a basis for the search algorithms considered in the book.

2.2 Search in Shadowed Space

Now let us consider an algorithm of search in shadowed space [29]. This algorithm follows the direction of the studies in shadowing and visibility [65, 66] problems in which it is assumed that the visibility of the target depends on the topography specified by certain continuous function.

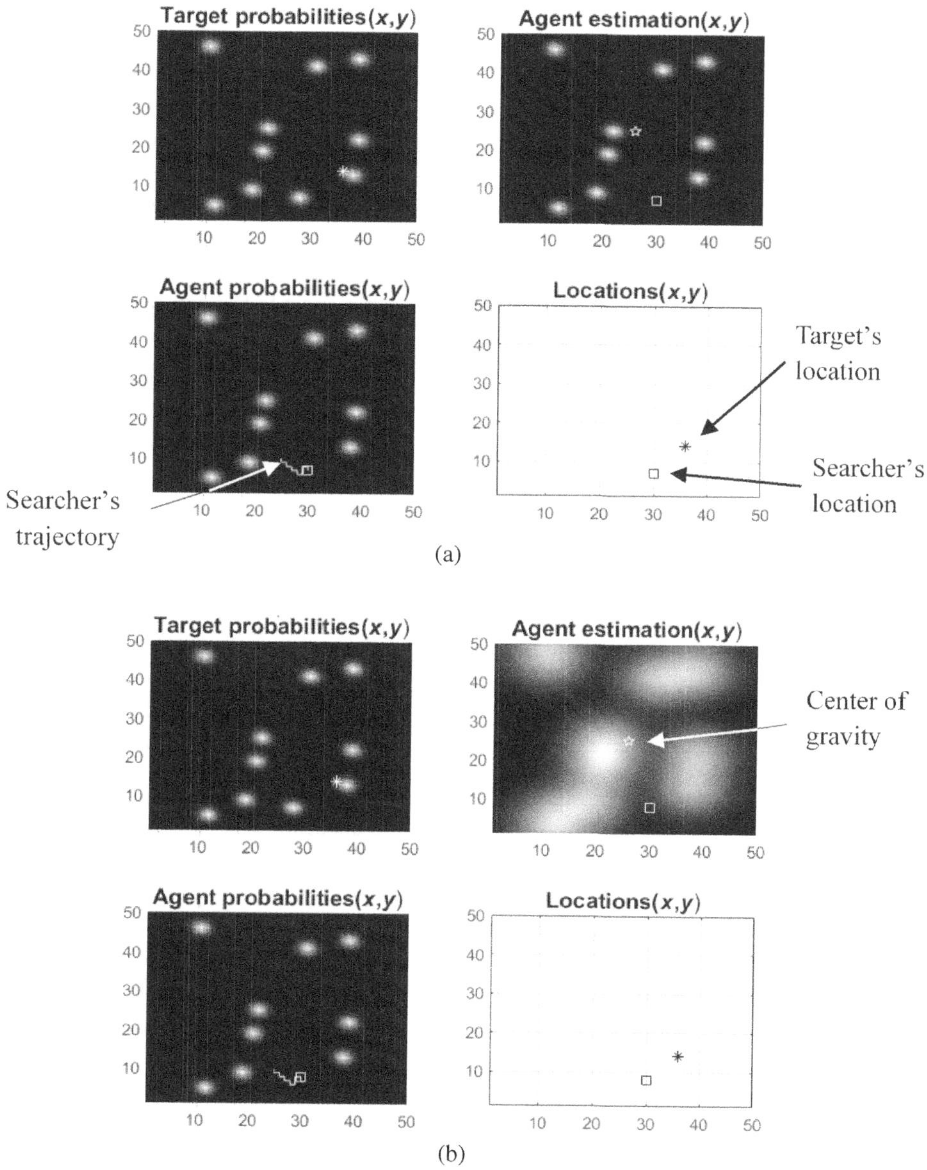

Figure 2.1 Actions of the Algorithm 2.1 and probabilities over the cells' set: (a) the state before estimating probabilities by Algorithm 2.3 and (b) the state after estimating probabilities.

In the algorithm it is assumed that similar to the Ohsumi model [45], the target moves according to a diffusion process and that, as above, detection is errorless.

The algorithm includes two stages: at the first stage the algorithm analyzes the visibility of the domain and creates a finite set of potential cells from which the searcher can observe a maximal part of the domain, and at the second stage the algorithm governs the online search in which the searcher moves over the set of potential cells. The goal is to find a path over a domain such that the target is detected in minimal expected time.

Let $C = \{c_1, c_2, \ldots, c_n\}$ be a set of cells and let $\mathcal{T} : C \to \mathbb{R}^+$ be a bounded function on C. For any observed area $O \subset C$, function $\mathcal{T}$ defines a measure $\#O$, which specifies the visibility of the area O.

Let c_1 and c_2, $c_1 \neq c_2$, be two cells from C and let O_1 be a set of cells, which are visible from cell c_1, and O_2 be a set of cells, which are visible from cell c_2.

If $c_1, c_2 \in (O_1 \cap O_2)$, then the cells c_1 and c_2 are called mutually comparable; otherwise, the cells are called mutually incomparable.

Assume that c_1 and c_2 are comparable. If $\#O_1 > \#O_2$, then it is said that the cell c_1 dominates on the cell c_2 (and c_2 is dominated by the cell c_1).

Potential cell is a cell that is not dominated by any of the cells, which are comparable with it.

Given two cells c_i and c_j, a visibility measure of the cell c_j with respect to the cell c_i is defined as follows:

$$\alpha_{ij} = \frac{\#(O_i \cap O_j)}{\#O_j}.$$

To check the comparability of the cells and their dominance and to control the number of potential cells, an overlap factor $a \in [0, 1]$ is applied. If $\alpha_{ij} > a$, then the observed area O_i includes much of the observed area O_j with respect to the factor a.

The visibility analysis algorithm identifies a set of potential cells with respect to a given overlap a with accuracy $\varepsilon \ll 1$. The algorithm starts with an initial value a_0 of overlap degree and chooses potential cells with respect to a_0. Then it increases the overlap value up to a_1 and chooses additional potential cells. The process continues up to the overlap a_n such that $0 \leq a_n - a \leq \varepsilon$. The algorithm is outlined as follows.

Algorithm 2.5 Visibility analysis

Input: Set $C = \{c_1, c_2, \ldots, c_n\}$, topography $\mathcal{T} : C \to \mathbb{R}^+$ and accuracy $\varepsilon \ll 1$.
Output: The set $\tilde{C} \subset D$ of potential cells.

Set $a = 0$: initial overlap value.
Set $l = 0$ and $u = 1$: low and upper bounds of a.
Do:
    Mark all cells $c \in C$ as non-dominated.
    Set $\tilde{C} = \emptyset$: initial potential cells set.
    For each cell $c_i \in C$, $i = 1, 2, \ldots, n$, do:
        For each cell c_j comparable with c_i do:
            Set $\alpha_{ij} = \#(O_i \cap O_j)/\#O_j$ and $\alpha_{ji} = \#(O_i \cap O_j)/\#O_i$
            If $\alpha_{ij} > a$, then
                Mark cell c_j as dominated (by c_i) with respect to a,
            Else if $\alpha_{ji} > a$, then
                Mark cell c_i as dominated (by c_j) with respect to a,
            Else
                Choose dominated cell at random.
            End if.
        End for.
    End for.
    Insert all non-dominated cells into the set $\tilde{C}$.
    If the observed areas of the cells from the set $\tilde{C}$ cover the set C, then
        Set $u = \alpha$.

21. Else
    Set $l = \alpha$.
22. End if.
23. Set $a = (l + u)/2$.
24. While observed areas of the cells from the set $\tilde{C}$ does not cover the set C or $(u - l) > \varepsilon$.
25. Return set $\tilde{C}$ of potential cells.

The algorithm runs over the cell set C and finds a subset $\tilde{C}$ of the cells, which should be visited by the search to find the target in one of the areas of its visibility. Note that all the cells from the subset $\tilde{C}$ are on the local maximums of the function $\mathcal{T}$.

The value a of overlap depends on the function $\mathcal{T}$. If the overlap value $a = 1.0$, then the algorithm returns the set $\tilde{C} = C$ and the searcher should move over all cells of the domain. However, numerical simulations demonstrated that for the reasonable topography $\mathcal{T}$, the overlap values $a \sim 0.95$ provide the best search results versus search time. Examples of the sets of potential cells are shown in Figure 2.2.

The obtained cells are used as intermediate cells of the searcher's path over the domain. Similar to the Algorithm 2.1, the algorithm governs the search agent toward the expected target's location using the probabilistic cost. In general, the algorithm acts as follows.

Assume that the search agent starts in the cell $c_0 = c(t_0)$. Then, to choose the best next cell, the agent evaluates the paths to all cells from the set $\tilde{C}$. The path to each cell $\tilde{c} \in \tilde{C}$ consists of a curve close to straight line from c_0 to $\tilde{c}$ and of cell $\tilde{c}$ itself, in which the searcher stays up to a time T.

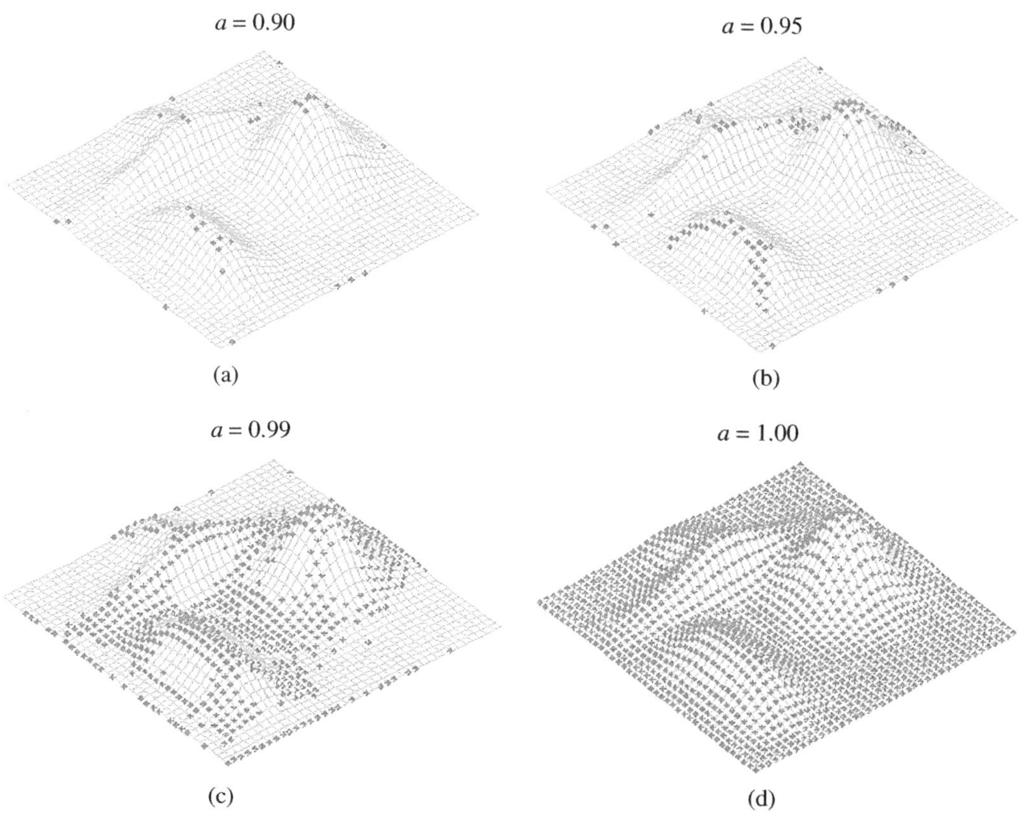

Figure 2.2 Potential cells obtained by Algorithm 2.5 with different values of overlap factor a.

The value T depends on the searcher's speed v and the size of the domain. During the period $(t_0, \tilde{t}]$ the search moves to the cell

$$c(t) = \frac{c_0 + vt(\tilde{c} - c_0)}{dist(\tilde{c}, c_0)}, t_0 < t \leq \tilde{t},$$

and during the period $(\tilde{t}, T]$ stays in the cell $c(t) = \tilde{c}, \tilde{t} < t \leq T$.

For the square grid of the size $n_1 \times n_2$, $n_1 = n_2$, in which coordinates of the cells are $\tilde{c} = (\tilde{j}_1, \tilde{j}_2)$ and $c_0 = (j_{01}, j_{02})$, the times $\tilde{t}$ and T are

$$\tilde{t} = \frac{|\tilde{j}_1 - j_{01}| + |\tilde{j}_2 - j_{02}|}{v} \text{ and } T = \frac{2n_1}{v}.$$

The choice of the next cell $\tilde{c}$ among the cell from the set $\tilde{C}$ is conducted using the following criterion. Let $P(\tilde{c}, T)$ be a probability of detecting the target and let $\tilde{T}(\tilde{c})$ be an expected time needed for detecting the target up to the time T while the searcher moves from its current position c_0 to the cell $\tilde{c}$ and then stays in it. In the simplest scenario, it can be assumed that the probability $P(\tilde{c}, T)$ is equal to the probability of target's location in the observed area $\tilde{O}$, which includes the cell $\tilde{c}$ and the time is $\tilde{T}(\tilde{c})$ equivalent to the number of steps need to move from the cell c_0 to the cell $\tilde{c}$. For general definitions of these values, see our report [29].

In the algorithm, it is assumed that the searcher chooses its next cell $\tilde{c}$ such that the relation $\frac{P(\tilde{c},T)}{\tilde{T}(\tilde{c})}$ reaches its maximum over the cells $\tilde{c} \in \tilde{C}$. The algorithm is outlined as follows.

Algorithm 2.6 Search in shadowed space

Input: Set $C = \{c_1, c_2, \ldots, c_n\}$, topography $\mathcal{T} : C \to \mathbb{R}^+$, set $\tilde{C} \subset C$ and time T.
Output: The cell $\tilde{c}$ from which the target is visible.

Set $t = 0$.
Start in the cell $\tilde{c}(t)$.
While target is not visible from $\tilde{c}(t)$ do:
  For each potential cell $\tilde{c} \in C$ do:
    Define probability $P(\tilde{c}, T)$ and time $\tilde{T}(\tilde{c})$.
    Set $\tilde{c}(t+1) = \text{argmax}_{\tilde{c} \subset \tilde{C}}\{P(\tilde{c}, T)/\tilde{T}(\tilde{c})\}$.
  End for.
  If $\tilde{c}(t+1) = \tilde{c}(t)$ then
    Stay in $\tilde{c}(t)$ during the average time of previous moves,
  Else
    Move to the cell $\tilde{c}(t+1)$.
    If during the movement to the cell $\tilde{c}(t+1)$ the target is observed, then
      Stop
    End if.
  End if.
  Set $t = t + 1$.
End while.
Return cell $\tilde{c}(t)$.

The algorithm acts online and terminates, resulting in a search path over a domain D and corresponding terrain topography $\mathcal{T}(y)$, $y \in D$. Snapshots of the searcher's motion over a domain and observed areas are shown in Figure 2.3.

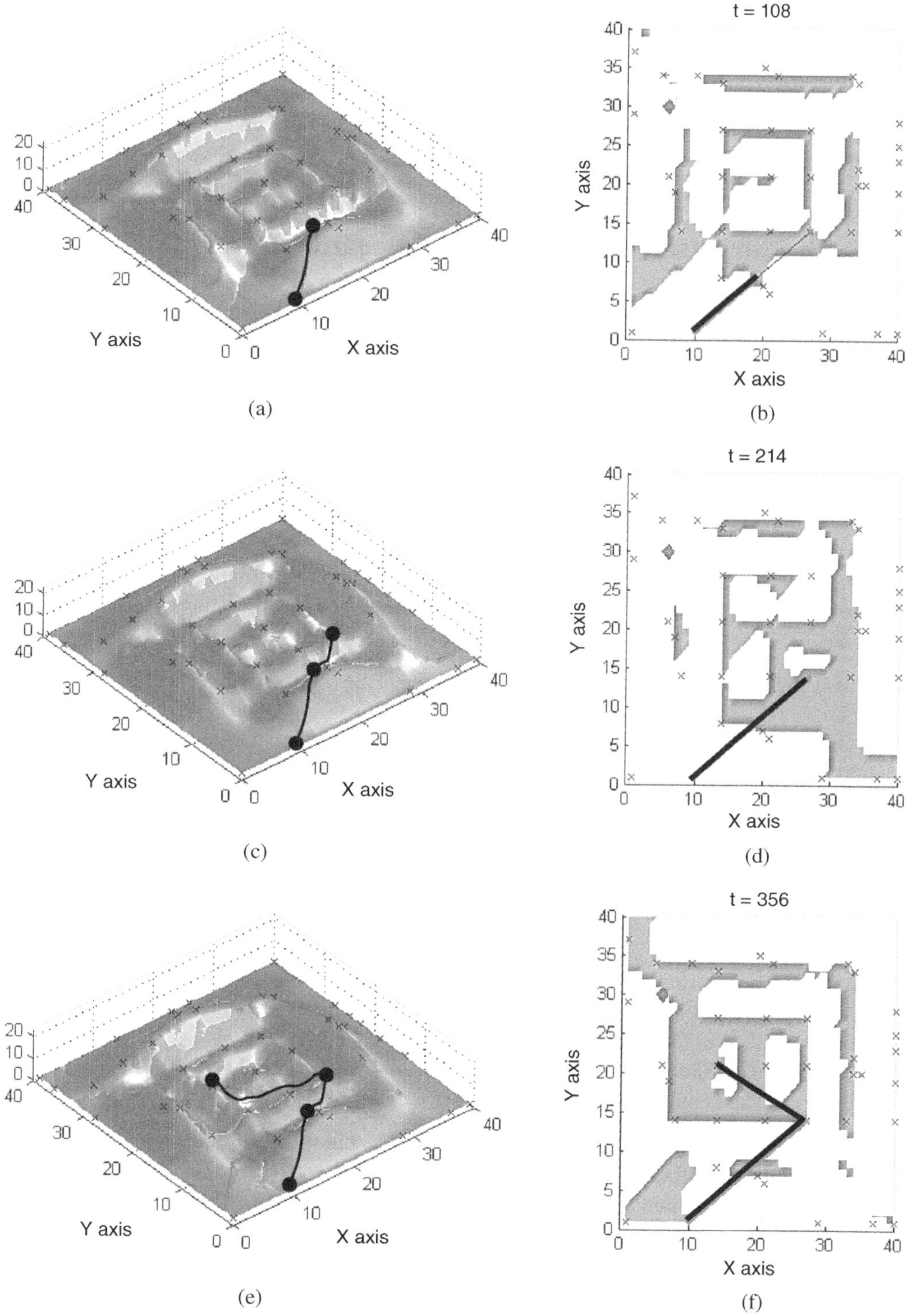

Figure 2.3 Snapshots of the searcher's motion over the domain governed by Algorithm 2.6: (a) and (b) – time $t = 108$, (c) and (d) – time $t = 214$, and (e) and (f) – time $t = 356$ at which the target was observed. Black curves in the figures (a), (c), and (e) depict the searcher's trajectory over the domain's topography and black lines in the figures (b), (d), and (f) depict the searcher's trajectory over a plane grid. The crosses $\times$ depict the potential cells, and gray areas in the figures (b), (d), and (f) show the observed areas from the searcher's locations.

Figure 2.3a, c, and e show the searcher's movements over the topography on the domain, and Figure 2.3b, d, and f show the movements of the searcher over the plane grid.

The algorithm effectively detects both static and moving targets. According to numerical simulations, in the search for static and moving targets, its only $3-10$ percent slower than the best result obtained by the brute force search. Along with that, it requires an offline visibility analysis of the terrain that decreases the usability of the algorithm.

2.3 Search with False-Positive and False-Negative Errors

Finally, let us consider an extension of the Algorithm 2.1 to the detection of the errors of the first and the second type [5]. This algorithm extends the algorithm of search for erroneous detections [34, 35].

As above, let $C = \{c_1, c_2, \ldots, c_n\}$ be a finite set of cells and $p_i(t)$, $i = 1, 2, \ldots, n$, are the probabilities of targets' location in the cells $c_i \in C$ at time $t = 0, 1, 2, \ldots$ The area observed by the search agent at time t is denoted by $O(t) \subset C$.

In contrast to Algorithm 2.1, which applies the detection function $\varphi(O(t)) = \mathbf{1}(c(t), O(t))$ of errorless detection, here it is assumed that the detection can be erroneous with the errors of the first and the second type.

Let $\varphi(O(t)) \in [0, 1]$ and $\psi(O(t)) \in [0, 1]$ be a detection and an error functions, respectively. An observation result $z(O(t))$ of the area $O(t)$ is defined as follows. Assume that the target is in the cell $c(t)$. Then

- if $c(t) \in O(t)$, then $z(O(t)) = 1$ with probability $\varphi(O(t))$ and $z(O(t)) = 0$ with probability $\overline{\varphi}(O(t)) = 1 - \varphi(O(t))$;
- if $c(t) \notin O(t)$, then $z(O(t)) = 1$ with probability $\psi(O(t))$ and $z(O(t)) = 0$ with probability $\overline{\psi}(O(t)) = 1 - \psi(O(t))$.

The expected location probabilities are denoted by $\tilde{p}_i(t)$; at initial time moment $t = 0$, these are equivalent to the location probabilities $p_i(t)$, $\tilde{p}_i(0) = p_i(0)$, $i = 1, 2, \ldots, n$. As above, the search process is conducted as follows [5]:

1) At each time t, the searcher observes an area $O(t) \subset X$, and obtains observation result $z(O(t))$. Given $z(O(t))$, probabilities $\varphi(O(t))$ and $\psi(O(t))$, and the number of targets k, the searcher decides whether to continue the search or to terminate.
2) If the searcher decides to continue the search, it defines the expected location probabilities $\tilde{p}_i(t+1)$, $i = 1, 2, \ldots, n$, for the next time $t+1$.
3) Given the expected probabilities $\tilde{p}_i(t+1)$, $i = 1, 2, \ldots, n$, the searcher chooses the next observed area $O(t+1) \subset X$ and continues search from step 1 with $t = t+1$.

The goal is to find a policy for choosing the areas $O(t) \subset X$ that results in finding the target with maximal probability in minimal number of steps.

The algorithm that implements this search process is based on the Algorithm 2.1. Expected location probabilities $\tilde{p}_i$ and $\tilde{\pi}_i$, $i = 1, 2, \ldots, n$, are calculated using the probabilities φ and ψ as follows:

$$\tilde{p}_i(t+1) = \frac{\varphi(\{c_i\})\psi(\{c_i\})\,\tilde{p}_i(t) + \varepsilon}{\sum_{j=1}^{n}(\varphi(\{c_j\})\psi(\{c_j\})\tilde{p}_j(t) + \varepsilon)}, \qquad (*)$$

$$\tilde{\pi}_i(t+1) = \frac{(1 - z(\{c_i\}))\tilde{\pi}_i(t)}{\sum_{j=1}^{n}(1 - z(\{c_j\}))\tilde{\pi}_j(t)}, \qquad (**)$$

where ε, $|\varepsilon| \ll 1$ is a uniformly distributed random variable. Such an updating of the probabilities $\tilde{p}_i$ corresponds to the random movement of the target.

The other values are calculated by the same procedures as in Algorithm 2.1. Namely, the number m of steps required for the estimations of the location probabilities by diffusion of the location probabilities is given by the formula (see line 3 in Algorithm 2.3)

$$m = \lceil I(\tilde{\pi}(t+1)) \rceil \cdot [\text{dist}(\tilde{c}, y)],$$

where

$$I(\tilde{\pi}(t+1)) = \log_s(n) - \sum_{i=1}^{n} \tilde{\pi}_i(t+1)\log_s(\tilde{\pi}_i(t+1))$$

is an information lack for an observed area of the size $s = O(t)$, and $\text{dist}(\tilde{c}, y)$ is the distance between expected target's location $\tilde{c}$, which is a center of distribution, and current searcher's location y.

Diffusion of the location probabilities is defined by the transitions matrix of Markov process, which governs the target's motion; as above, probabilities governed by the diffusion are denoted by $\tilde{\mathcal{p}}_i$, $i = 1, 2, \ldots, n$.

The choice of the next observed area $O(t+1)$ is governed by the decision-making procedure, which follows the expected probabilities defined by the equation (*) and (**). The procedure is outlined as follows (in parallel to Algorithm 2.2).

Algorithm 2.7 Function next observed area with uncertain detections

---

Input: Array of expected probabilities $\tilde{\mathcal{p}} = (\tilde{\mathcal{p}}_1, \tilde{\mathcal{p}}_2, \ldots, \tilde{\mathcal{p}}_n)$, set of movements $\mathbb{A}$, center $\tilde{c}$ of distribution $\tilde{\mathcal{p}}$, and current searcher's location y.

Output: Area $O \subset C$.

---

1. For all movements $\mathbb{a}_k \in \mathbb{A}$ do:
2. Determine the area O_k according to the movement $\mathbb{a}_k$.
3. Set $\mathcal{w}_k = \sum_{c_j \in O_k} \tilde{\mathcal{p}}_j$.
4. Calculate probabilities α_{ki} by equation (*) assuming that $c_i \notin O_k$.
5. Calculate probabilities β_{ki} by equation (*) assuming that $c_i \in O_k$.
6. Set $d(O_k) = \mathcal{w}_k \text{dist}(\alpha_k, \tilde{\mathcal{p}}) + (1 - \mathcal{w}_k) dist(\beta_k, \tilde{\mathcal{p}})$.
7. End for.
8. Set $O = \text{argmax}\{d(O_k)$ over all $O_k\}$; ties are broken randomly.
9. Return O.

The algorithm of search with erroneous detections follows the same steps of Algorithm 2.1 in which probabilities are updated by the equations (*) and (**), and the next observed areas are chosen by Algorithm 2.7. The algorithm is outlined as follows.

Algorithm 2.8 Search with erroneous detections

Input: Set $C = \{c_1, c_2, \ldots, c_n\}$, probabilities $p_i(0)$, $i = 1, 2, \ldots, n$, matrix $\rho = [\rho_{ij}]_{n \times n}$, and the set $\mathbb{A}$ of possible movements.

Output: Area $O(t) \subset C$ such that $\varphi(O(t)) = 1$.

---

1. Set $t = 0$.
2. Set $\tilde{p}_i(t) = p_i(0)$, $i = 1, 2, \ldots, n$.

3. Choose $O(t)$ arbitrary.
4. While $z(O(t)) \neq 1$, do
5. Calculate probabilities $\tilde{p}_i(t+1)$ by equation (*), $i = 1, 2, \dots, n$.
6. Calculate probabilities $\tilde{\pi}_i(t+1)$ by equation (**), $i = 1, 2, \dots, n$.
7. Calculate weights $w_k = \sum_{c_j \in O_k} \tilde{\wp}_j$ for the movements $a_k \in \mathbb{A}$.
8. If $w_k = 0$ for each k then
9. Set estimated probabilities $\tilde{\wp}$ by Algorithm 2.3,
10. Else
11. Set $\tilde{\wp} = \tilde{p}(t+1)$,
12. End if.
13. Calculate probabilities α_{ki} by equation (*) assuming that $c_i \notin O_k$.
14. If $\tilde{\wp}d(\tilde{c}, y_k) > w_k \text{dist}(\alpha_k, \tilde{p}(t+1))$ for at least one k
15. Set $\tilde{\pi}_i(t+1) = \tilde{p}_i(t+1)$, $i = 1, 2, \dots, n$.
16. End if.
17. Choose the next observed area $O(t+1)$ by Algorithm 2.7.
18. Move to the area $O(t+1)$.
19. Set $t = t+1$.
20. End while.
21. Return: the target is observed in the area $O(t)$.

The snapshot of the algorithm's activity is shown in Figure 2.4. In the figure, detection probabilities are $\varphi(O(t)) = 0.9$ and probabilities of false-positive detections are $\psi(O(t)) = 0.01$ for each time t, and the value ε is drawn from the interval $[-0.001, 0.001]$.

Simulations of the Algorithm 2.8 demonstrate that it terminates with finding the target in more than 95% of trials. Note that since the searcher moves in the observed areas and observes the cells in the areas during the motion, percentage of successful trials is greater than the 90% of correct detections. Percentage of successful trials increases with the size of observed areas up to 99.7% while each area covers the domain. However, for the same reason percentage of false-positive detections also increases.

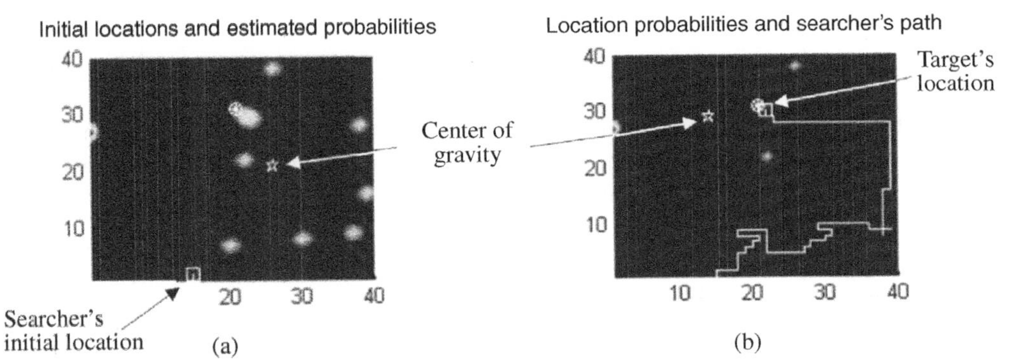

Figure 2.4 Actions of the Algorithm 2.8 and probabilities over the cells' set: (a) initial locations of the target and the searcher and estimated probabilities; (b) target's location probabilities and trajectory of the searcher at the end of the search.

2.4 Conclusions

In this chapter, we considered three simple algorithms of search for static and moving target in the gridded domain. These algorithms follow general approach of searching Markovian target [6] by observation of certain areas of the domain.

Algorithm 2.1 governs the search agent with errorless detection abilities and terminates when the agent recognizes the target in its close neighborhood. The main issue that was solved in this algorithm is making a decision about the next step, while the probabilities of finding the target in the observed areas are zero.

To resolve this uncertainty, the algorithm considers the expected probabilities of finding the target. Namely, it emulates the searcher waiting for several steps, while the target continues its motion and then makes a decision using expected probabilities of finding the target. After making the decision, the search returns to the real location probabilities and continues the search.

Algorithm 2.6 governs the searcher in the domain with limited visibility. Similar to the previous algorithm, the search has errorless detection abilities, but the observed areas are defined by a topography function and depend on the searcher's location. In such a scenario, the previous algorithm also terminates by finding the target, but its effectiveness is insufficient.

To resolve the uncertainty in visibility of the areas, the algorithm prior conducts a visibility analysis of the domain and then governs the agent over the cells which provide good enough visibility of the domain. During the motion, the searcher moves to the cell which is closest to the searcher's current location and provides maximal probability of seeing the target.

Finally, Algorithm 2.8 governs the searcher with detection errors of the first and the second types. This algorithm extends Algorithm 2.1 and applies the same principles of resolving uncertainties in the choices of the next observed areas.

To resolve the uncertainty led by the detection errors, for each observed area the algorithm considers three values: observation result, detection probability, and detection error. The first value is used for terminating the algorithm and updating the expected probabilities, and two other probabilities are used for resolving uncertainties in the choice of the next observed areas.

The considered algorithms have a common procedure – prediction of the searcher's location, which is based on the rule of the target's motion; in the case of static target, in the prediction stage it is assumed that the target is Brownian, while its real location probabilities remained constant. The history of search is stored in the form of updated target's location probabilities.

Note that each of these algorithms can be directly extended to the search by a group of searchers for several targets. In such a scenario, the searchers deal with common arrays of targets' location probabilities, while prediction and decision-making are conducted separately by each searcher.

3

Problem Formulation and Basic Search Procedure

In this book, we consider the problem of search and detection of multiple targets by a group of mobile search agents acting in the gridded domain represented by the set of cells.

The agents can be considered as mobile robots that are equipped with sensors and obtain information about the target's location with false positive and false negative sensing errors. In other words, the agent can erroneously avoid an existing target and can erroneously recognize the target in the empty cell.

The agents obtain information about the targets' locations separately but can share information among the group members. Information about the targets' locations is stored in the form of probability maps that are available to all agents.

Note that such an approach does not require the existence of a central station; the maps can be stored in the on-board computer of each agent and can be updated with respect to the sensed and shared information.

The goal is to govern the search agents such that all targets will be detected in a minimal number of steps.

3.1 Basic Assumptions

The considered algorithms of search follow general principles of the search and detection theory [55–57, 63] and implement probabilistic and informational methods of tracking targets under uncertainty [35, 36]. They continue the direction of the algorithms of search considered in Chapter 2 and consider the search for Markovian targets [6] following the occupancy grid techniques [12, 13].

Integration of the data obtained by different sensors and different agents follows general Bayesian scheme. We consider both the sensors with constant sensitivities and sensitivities varying with respect to the status of the search.

As indicated above, in the search by a group of the search agents, the agents share information and store history of the search by updating the targets' location probabilities. We assume that the searchers manage three types of probability maps that represent the location probabilities:

- Global probability map includes target's location probabilities that are defined by fusion of the information obtained by the group members; this map is available to all searchers in the group.
- Local probability map includes the target's location probabilities that are obtained by fusion of the information obtained by the on-board sensors of the agent; this map is available to the agent.
- Sensor probability map includes the target's location probabilities that are obtained by a single sensor available to a searcher.

Let $C = \{c_1, c_2, \ldots, c_n\}$ be a finite set of cells and consider a set of m, $1 \leq m \ll n$, search agents searching for hidden targets in the set C. It is assumed that the searcher and the target can occupy only a single cell.

The state of a cell c_i, $i = 1, 2, \ldots, n$, is specified by the value $s_i = s(c_i) \in \{0, 1\}$ such that $s_i = 0$ means that the cell c_i does not contain a target and $s_i = 1$ means that the cell c_i contains a target. In case we need to stress the time t of the sensing, we will use the notation $s_i^t = s\left(c_i, t\right)$; otherwise, we omit it.

Note that at any time t for each cell c_i the probabilities of the events $\{s_i = 0\}$ and $\{s_i = 1\}$ are mutually exclusive

$$Pr\left\{s_i = 0\right\} + Pr\left\{s_i = 1\right\} = 1.$$

Each jth search agent is equipped with a variety of sensors $\mathbb{s}_{jk}$, $k = 1, 2, \ldots l$, that provide not necessarily correct information about the states s_i of the cell c_i, $i = 1, 2, \ldots, n$. The probability of target's detection in the cell c_i depends on the distance $dist(c_i, y)$ between this cell and the searcher's location y and is defined by the Koopman random search formula [39]

$$Pr\left\{\text{target detected in } c_i \mid \text{target located in } c_i\right\} = 1 - \exp\left[-\kappa\left(dist\left(c_i, y\right), \tau\right)\right],$$

where $\kappa(dist(c_i, y), \tau)$ represents the search effort applied to the cell c_i given the distance $dist(c_i, y)$ and observation period τ. It is assumed that the detection probability is higher as the distance $dist(c_i, y)$ is shorter and as the period τ is longer.

To formalize the possibility of both false positive and false negative detection errors, we assume that there are both true and dummy targets and the targets of both types broadcast the signals indicating their presence in the cells. The signals sent by the true targets are considered as true alarms and the signals sent by the dummy targets are considered as false alarms that represent false-positive errors.

Following the Koopman random search formula, the probability that the alarm perceived is defined as follows:

$$Pr\left\{\text{alarm percieved} \mid \text{alarm sent}\right\} = \exp\left[-dist\left(c_i, y\right) / \lambda_{jk}\right],$$

where $\lambda_{jk} = \lambda\left(\mathbb{s}_{jk}\right)$ is the sensitivity of the sensor $\mathbb{s}_{jk}$ of the jth agent.

The probabilities

$$p_i = p\left(c_i\right) = Pr\left\{\text{alarm sent from } c_i\right\}$$

of sending alarms from cells $c_i \in C$, $i = 1, 2, \ldots, n$, are defined as targets' location probabilities and are represented by the appropriate probability map.

As indicated above, the searchers can share information about the targets' locations. Such activity of the searchers is outlined as follows.

1) The searchers start with the probability map

   $$P(t) = \left\{p_1(t), p_2(t), \ldots, p_n(t)\right\}$$

   that defines initial probabilities $p_i(t)$ of detecting the targets in cells $c_i \in C$, $i = 1, 2, \ldots, n$, at time t.
2) The searchers receive true or false alarms from the cells in which the targets can be located.
3) Following the received alarms, each searcher j updates the probability map for each kth sensor, $k = 1, 2, \ldots, l$,

   $$P^{\text{sensor}}(j, k, t) = \left\{p^{\text{sensor}}_{s_1=1}(j, k, t), p^{\text{sensor}}_{s_2=1}(j, k, t), \ldots, p^{\text{sensor}}_{s_n=1}(j, k, t)\right\}$$

where

$$p^{\text{sensor}}_{s_i=1}(j,k,t) = Pr\left\{\text{sensor } \mathbb{s}_{jk} \text{ identifies a true target in } c_i \text{ at time } t\right\}.$$

4) The resulting sensor probability maps $P^{\text{sensor}}(j,k,t)$, $k = 1, 2, \ldots, l$, are combined into the agent's probability map

$$P^{\text{agent}}(j,t) = \left\{p^{\text{agent}}_{s_1=1}(j,t), p^{\text{agent}}_{s_2=1}(j,t), \ldots, p^{\text{agent}}_{s_n=1}(j,t)\right\}$$

where probabilities $p^{\text{agent}}_{s_i=1}(j,t)$ combine the sensor probabilities and represent target's locations in the cells c_i, $i = 1, 2, \ldots, n$, from the agent's point of view.

5) Finally, global probability map

$$P^{\text{global}}(t) = \left\{p^{\text{global}}_{s_1=1}(t), p^{\text{global}}_{s_2=1}(t), \ldots, p^{\text{global}}_{s_n=1}(t)\right\}$$

defines the probabilities $p^{\text{global}}_{s_i=1}(t)$ combine the agents' probabilities and represent target's locations in the cells c_i, $i = 1, 2, \ldots, n$, as they are known to the group of agents.

The probability maps at the levels of sensors and of agents are fused using Bayesian scheme that allows direct calculations of the probabilities at each time t, given the probabilities at the previous time $t-1$ and the information obtained by the sensors at time t. At the initial time $t = 0$, the probabilities are defined based on topographic data and prior information or, the worst case, are specified by a uniform distribution on the grid.

General scenario of the targets' detection by a group of mobile agents follows the lines of the scenario presented in Section 2.3 and is outlined as follows.

1) The search agent observes the available cells by its sensors, obtains true or false information about the target's location, and updates the probability maps.
2) Using the updated probability maps, the search agent makes a decision regarding its next movement. Decision is based on local or global probability maps.
3) Following the decision, the agent makes a movement in the chosen direction and continues the search up to detecting all targets or up to predefined time.

The goal is to define the strategy of the search agents such that they find all targets in minimal time.

In the terms of the searchers' trajectories (see e.g. [4]), the goal is defined as follows. Denote by

$$\tau_j(T) = \left(c_j(0), c_j(1), \ldots, c_j(T)\right)$$

trajectory of the jth agent starting from its initial cell $c_j(0)$ and moving to the cell $c_j(t)$.

Being in the cell $c_j(t)$, the agent makes a decision regarding its next location following certain policy $\pi_j(P)$ that prescribes how to choose the next cell $c_j(t+1)$ given a probability map P. For simplicity assume that policy $\pi_j(P)$ for each agent does not depend on the time and for any t depends only on the applied probability map P. Application of the policy $\pi_j(P)$ results in the action $\mathbb{a}(t)$ that controls the agent's move from the current cell $c_j(t)$ to the next cell $c_j(t+1)$.

Denote by $T_\theta(p \mid \pi_1(P), \pi_2(P), \ldots, \pi_m(P))$ the time required to detect the θth target, $\theta = 1, 2, \ldots, \xi$, with probability p given the agents' policies $\pi_1(P), \pi_2(P), \ldots, \pi_m(P)$. Then the goal is to define such policies

$$\left(\pi_1^*(P), \pi_2^*(P), \ldots, \pi_m^*(P)\right) = \underset{\left(\pi_1(P),\pi_2(P),\ldots,\pi_m(P)\right)}{\text{argmin}} \; \max_{\theta=1,2,\ldots,\xi} T_\theta(p)$$

that result detecting the last ξth target in minimal time.

Since this problem is computationally hard and cannot be solved exactly for practically interesting situations, we consider several heuristic methods and apply deep Q-learning techniques.

3.2 Sensor's Fusion and Dynamic Probability Map

As indicated above, it is assumed that each agent is equipped with several sensors $\mathbb{s}_{jk}$, $k = 1, 2, \ldots l$, that independently provide, not necessarily accurate, information about the cells' states s_i, $i = 1, 2, \ldots, n$. Fusion of the information obtained by the sensors is conducted as follows.

Let $\mathbb{s}_{j1}$ and $\mathbb{s}_{j2}$ be two independent sensors installed on the jth agent and let $\tilde{s}_{j1}(c_i, t)$ and $\tilde{s}_{j2}(c_i, t)$ be the signals obtained by these sensors at time t. Then, the probability that the target is located in the cell c_i, which is the state $s_i(t) = 1$, is defined as follows:

$$Pr\left\{s_i(t) = 1 \mid \tilde{s}_{j1}(c_i, t) = 1, \tilde{s}_{j2}(c_i, t) = 1\right\} = \frac{Pr\left\{\tilde{s}_{j2}(c_i, t) = 1 \mid s_i(t) = 1\right\} \times Pr\left\{s_i(t) = 1 \mid \tilde{s}_{j1}(c_i, t) = 1\right\}}{\sum_{s_i(t)} Pr\left\{\tilde{s}_{j2}(c_i, t) = 1 \mid s_i(t)\right\} \times Pr\left\{s_i(t) \mid \tilde{s}_{j1}(c_i, t) = 1\right\}},$$

where sum is taken over all possible values of $s_i(t)$; in the considered case, these values are $s_i(t) \in \{0, 1\}$.

Extension of this equation to l onboard sensors of the jth results in the probabilities

$$p_{s_i=1}^{\text{agent}}(j, t) = \frac{\prod_{k=1}^{l} p_{s_i=1}^{\text{sensor}}(j, k, t)}{\prod_{k=1}^{l} p_{s_i=1}^{\text{sensor}}(j, k, t) + \prod_{k=1}^{l}\left(1 - p_{s_i=1}^{\text{sensor}}(j, k, t)\right)}$$

of the targets' locations in the cells c_i, $i = 1, 2, \ldots, n$, as determined by the jth using its sensors. This equation is based on the approach known as "independent opinion pool" [13] where it is assumed that the sensors are conditionally independent and that their reliabilities and accuracies are equivalent.

Similarly, location probabilities of different agents can be fused to global probabilities

$$p_{s_i=1}^{\text{global}}(t) = \frac{\prod_{j=1}^{m} p_{s_i=1}^{\text{agent}}(j, t)}{\prod_{j=1}^{m} p_{s_i=1}^{\text{agent}}(j, t) + \prod_{j=1}^{m}\left(1 - p_{s_i=1}^{\text{agent}}(j, t)\right)}$$

of the targets' locations in cells c_i, $i = 1, 2, \ldots, n$, as determined by the group of the agents.

Note that in practical implementations, the use of global probabilities map requires a central unit that receives data from each agent and conducts the required computations.

The presented equations use the probabilities $p_{s_i=1}^{\text{sensor}}(j, k, t)$ of the targets' locations in the cells c_i, $i = 1, 2, \ldots, n$, as determined by sensors $\mathbb{s}_{jk}$, $k = 1, 2, \ldots l$, installed on agents, $j = 1, 2, \ldots, m$. These probabilities form a sensor probability maps $P^{\text{sensor}}(j, k, t)$ that are updated as follows.

At the initial time $t = 0$, the probabilities $p_{s_i=1}^{\text{sensor}}(j, k, t)$ are specified with respect to some initial distribution; if no information is available, these probabilities can be drawn by a uniform distribution. Then, these probabilities are updated as follows.

Let $\tilde{s}_{jk}(c_i, t)$ be a signal obtained about cell c_i relying on sensor $\mathbb{s}_{jk}$ of the jth agent at time t. Recall that in the considered scenario, $\tilde{s}_{jk}(c_i, t) = 1$ implies that cell c_i is occupied by a target and $\tilde{s}_{jk}(c_i, t) = 0$ implies that cell c_i is empty, both based on sensor $\mathbb{s}_{jk}$.

Then, the state probabilities of cell c_i updated by the sensor outputs are the following:

- if the percieved signal is $\tilde{s}_{jk}(c_i, t) = 1$, then assuming that the target was located at the cell c_i at time $t - 1$, the true positive probability is

$$p_{s_i=1,1}^{\text{sensor}}(j, k, t) = Pr\left\{s_i(t) = 1 \mid \tilde{s}_{jk}(c_i, t) = 1\right\} = \frac{Pr\left\{s_i(t-1) = 1\right\} \times Pr\left\{\tilde{s}_{jk}(c_i, t) = 1 \mid s_i(t) = 1\right\}}{\sum_{s_i(t)} Pr\left\{s_i(t-1)\right\} \times Pr\left\{\tilde{s}_{jk}(c_i, t) = 1 \mid s_i(t)\right\}};$$

- otherwise the percieved signal is $\tilde{s}_{jk}\left(c_i, t\right) = 0$ and the false-positive probability is

$$p^{\text{sensor}}_{s_i=1,0}(j,k,t) = Pr\left\{s_i(t) = 1 \mid \tilde{s}_{jk}\left(c_i,t\right) = 0\right\}$$
$$= \frac{Pr\left\{s_i(t-1) = 1\right\} \times Pr\left\{\tilde{s}_{jk}\left(c_i,t\right) = 0 \mid s_i(t) = 1\right\}}{\sum_{s_i(t)} Pr\left\{s_i(t-1)\right\} \times Pr\left\{\tilde{s}_{jk}\left(c_i,t\right) = 0 \mid s_i(t)\right\}}.$$

These equations define an updating scheme of the probabilities map for the sensor given new observations. They include the probabilities $Pr\left\{\tilde{s}_{jk}\left(c_i,t\right) = 1 \mid s_i(t) = 1\right\}$ that the sensor perceives the signal given that the target is in the cell c_i and the probability $Pr\left\{\tilde{s}_{jk}\left(c_i,t\right) = 0 \mid s_i(t) = 1\right\}$ that the sensor does not perceive a signal from cell c_i given that the target is in that cell.

To define these probabilities, denote by $\tilde{a}\left(c_i,t\right)$ an alarm that is sent from the cell c_i at time t. The value $\tilde{a}\left(c_i,t\right) = 1$ implies, truly or not, that the cell is occupied and the value $\tilde{a}\left(c_i,t\right) = 0$ implies, truly or not, that the cell is empty. Then, using the Koopman formula we obtain

$$Pr\left\{\tilde{s}_{jk}\left(c_i,t\right) = 1 \mid s_i(t) = 1\right\} = Pr\left\{\tilde{a}\left(c_i,t\right) = 1 \mid s_i(t) = 1\right\} \exp\left[-d\left(c_i,y_j\right)/\lambda_{jk}\right]$$

and consequently

$$Pr\left\{\tilde{s}_{jk}\left(c_i,t\right) = 1 \mid s_i(t) = 0\right\} = 1 - Pr\left\{\tilde{s}_{jk}\left(c_i,t\right) = 1 \mid s_i(t) = 1\right\},$$

where, as above, dist(c_i, y_j) is the distance between the cell c_i and the agent's location y_j and $\lambda_{jk} = \lambda\left(\mathbb{s}_{jk}\right)$ is the sensitivity of the sensor $\mathbb{s}_{jk}$ installed on the agent.

These equations enable calculating the occupation probabilities at each time t, given the probabilities at the previous time $t-1$ and the information obtained by the sensors at time t. The defined sensor fusion process is illustrated in Figure 3.1.

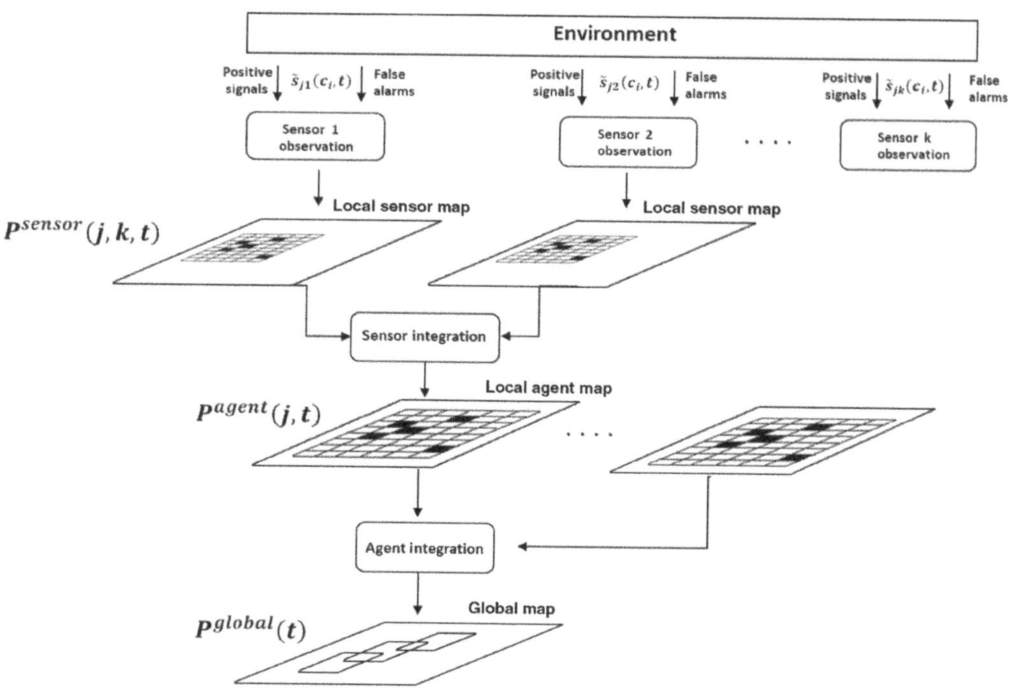

Figure 3.1 The process of sensor fusion.

The sensors receive signals $\tilde{s}_{jk}\left(c_i, t\right)$ from the environment. Part of these signals are positive signals from the targets, indicating the real locations of the targets, and the others are false alarms that correspond to false locations of the targets.

Based on the received signals, for each sensor a local sensor map is created. Then, each agent integrates sensor maps into a local agent map. Finally, a global map is created by combining the agent maps.

Such a hierarchical structure allows to consider the maps of each level separately and, consequently, to define more effective calculations, which uses only the required maps.

3.3 Search Policy and Sensing

The problem can be cosireder in two directions. In the search scenario, it is assumed that the target is found if at least one agent reached its location. In other words, the problem is considered as the path-planning problem widely accepted in the considerations of search for moving targets.

In the detection scenario, it is not required to arrive at the targets' locations, and it is assumed that the target is found if it is observed with some level of certainty. Such formulation follows classical considerations of the search and screening problem, resulting in the distribution of search efforts over the domain.

Below we concentrate on the first scenario.

3.3.1 Search Policy

In the search by the group of the search agents, the search policies of each agent depend on the activities of the other agents and on information obtained from them. Let us consider the step of the search process including communication between the agents.

In the search scenario, the step of the search process is outlined as follows.

At time t, the jth search agent is located in the cell $y_j(t)$ and percepts true and false alarms from the cells where the targets can be located. The quality of perception depends on the sensitivities λ of the agent's sensors and the distances $\mathrm{dist}(c_i, y_j(t))$ between the agent's cell $y_j(t)$ and the cells c_i, from which the alarms are sent.

After receiving the signals, the sensor probability maps $p^{\text{sensor}}_{s(c_i)=1}(j, k, t)$, $i = 1, 2, \ldots, n$, are updated, and the resulting sensor probability maps are combined into the agent's probability map $p^{\text{agent}}_{s(c_i)=1}(j, t)$, $i = 1, 2, \ldots, n$.

At this point, there are three possible scenarios of the searchers' activity with respect to communication:

1) If the agents act independently without communication, then the further decision about the next step is obtained basing on the agent's probability map $p^{\text{agent}}_{s(c_i)=1}(j, t)$, $i = 1, 2, \ldots, n$.
2) If the agents communicate and share complete information, then each agent creates a global probability map $p^{\text{global}}_{s(c_i)=1}(t)$, $i = 1, 2, \ldots, n$, and makes a decision based on this map.
3) Otherwise, if the agents share incomplete information, then following partial maps obtained from the other agents, each agent creates local probability map and makes a decision using this map.

Let us start with the scenario in which the agents share complete information. In this case, at each time t each agent is aware about its location $y_j(t)$, the probability map $p^{\text{agent}}_{s(c_i)=1}(j, t)$, and the probability map $p^{\text{global}}_{s(c_i)=1}(t)$, $i = 1, 2, \ldots, n$.

Assume that the search agent should move toward the highest probability $p^{\text{global}}_{s(c_i)=1}(t)$, $i = 1, 2, \ldots, n$, appearing in the global probability map. Since the agents' movements directly define the length of the trajectory, the agents' choice should minimize this length. Then, using the same criterion as in Algorithm 2.6, the next searcher's cell is chosen as follows:

$$y_j(t+1) = \underset{i=1,\ldots,n}{\operatorname{argmax}} \left\{ p^{\text{global}}_{s(c_i)=1}(t) / \text{dist}\left(c_i, y_j(t)\right) \right\}, \tag{*}$$

where $\text{dist}(c_i, y_j(t))$ is the distance between the agent's cell $y_j(t)$ and the cells c_i, from which the alarms were sent.

Using a global probability map with a reasonable search policy provides the best results in terms of minimal search time than partial or individual probability maps. However, the use of global probability map requires either transferring all available information to a central station and then broadcasting it to the agents or transferring all information to each agent and processing it by the on-board computer.

To decrease the quantity of the transferred information and of the computations, partial or individual maps can be used. In the first case, it is assumed that the search agent shares only those positions of the cells in which the probabilities of detecting the targets are relatively high. Such technique allows decreasing uncertainty in target locations and excluding some of the false alarms.

Formally, the sharing is implemented as follows. The data are transferred among the sensors $\mathbb{s}^k$ of the same type installed on different agents $j = 1, 2, \ldots, m$, and for this type, k, of sensors, the threshold probability P^*_k is specified. Over the agents, maximal probability

$$p^{\max}_{s(c_i)=1}(k,t) = \max_{j=1,\ldots,m} \left\{ p^{\text{sensor}}_{s(c_i)=1}(j,k,t) \right\}$$

is defined and if $p^{\max}_{s(c_i)=1}(k,t) > P^*_k$, then for each agent the probability $p^{\text{sensor}}_{s(c_i)=1}(j,k,t)$ of the sensor of the type k is updated as

$$p^{\text{sensor}}_{s(c_i)=1}(j,k,t) = \frac{p^{\max}_{s(c_i)=1}(k,t) \cdot p^{\text{sensor}}_{s(c_i)=1}(j,k,t)}{p^{\max}_{s(c_i)=1}(k,t) \cdot p^{\text{sensor}}_{s(c_i)=1}(j,k,t) + \left(1 - p^{\max}_{s(c_i)=1}(k,t)\right) \cdot \left(1 - p^{\text{sensor}}_{s(c_i)=1}(j,k,t)\right)}.$$

Such partial data sharing improves the agents' probability maps that allow better decisions even using the simple rule (*) for choosing next searchers' cells $y_j(t+1)$, $j = 1, 2, \ldots, m$.

The same decision rule (*) is applied to the agent's probability map $p^{\text{agent}}_{s(c_i)=1}(j,t)$, $i = 1, 2, \ldots, n$, which are created independently by each agent and do not require communication between the agents.

3.3.2 Search Time and Sensors' Sensitivity

In the considerations above, it was assumed that the sensitivity of the sensors is constant during the search and does not depend on the location of the searcher. Such assumption is rather strong and does not take into account specific conditions of search. One type of such conditions – the shadowed domain – was considered in Algorithm 2.6. Here we are interested in the dependence of search time T on the sensors' sensitivity $\lambda(\mathbb{s})$.

It is clear that more sensitive sensors enable detection of more targets at greater distances and shorten the time of search. On the other hand, more sensitive sensors perceive more information and so have greater probability of the errors of the second type. Thus, for each search mission it is required to define the sensitivity of the sensors which satisfies the requirements of search.

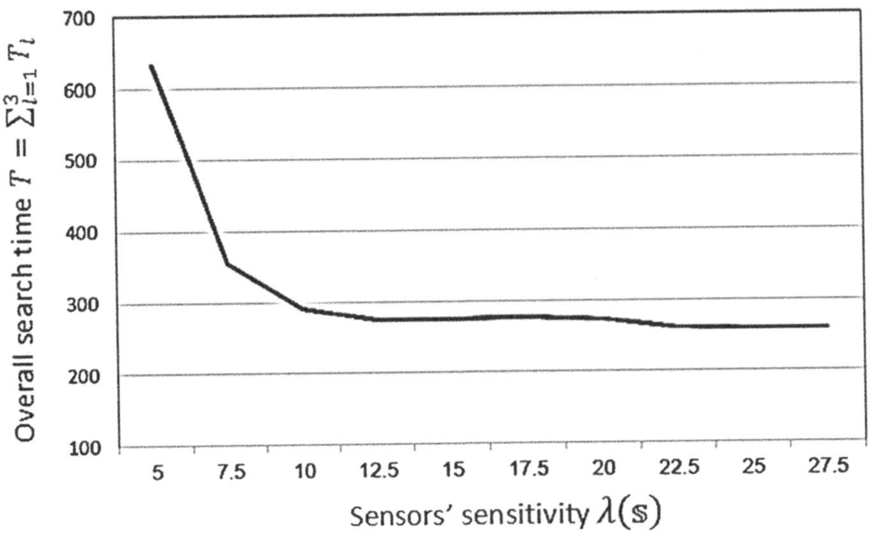

Figure 3.2 Dependence of search time on the sensors' sensitivity in the search for three targets by a single agent.

In Figure 3.2, we show the dependence of the search time on the sensitivity of the sensors obtained in numerical simulations of search for three targets by a single agent.

It is seen that the search time T decreases exponentially with the sensors' sensitivity $\lambda(\mathbb{s})$ such that for the values $\lambda(\mathbb{s}) > 10$, the increase in sensitivity has a minimal influence on the search time. That allows choosing the sensors with the sensitivity $\lambda(\mathbb{s}) \sim 10$ without loss of search efficiency.

Finally, there can be suggested a scenario, in which the sensitivity of the sensors varies during the search with respect to the updates of the probability maps. For example, the sensitivity updating can be conducted as follows:

$$\lambda(\mathbb{s}, t+1) = \begin{cases} \alpha\, \lambda(\mathbb{s}, t) & \text{if } p^{\max}_{s(c_i)=1}(k,t) \le P^k_l, \\ \lambda(\mathbb{s}, t) & \text{if } P^k_l \le p^{\max}_{s(c_i)=1}(k,t) \le P^k_u, \\ \frac{1}{\alpha}\lambda(\mathbb{s}, t) & \text{if } p^{\max}_{s(c_i)=1}(k,t) \ge P^k_u, \end{cases}$$

where $\lambda(\mathbb{s}, t)$ and $\lambda(\mathbb{s}, t+1)$ stand for the sensitivity of the sensors at time t and $t+1$, respectively, $p^{\max}_{s(c_i)=1}(k,t)$ is a maximal probability over the agents of detecting the target at time t by the sensor of type k, P^k_u and by P^k_l are upper and lower threshold probabilities and $\alpha \ge 1$ is an updating coefficient.

3.3.3 Numerical Simulations

Numerical simulations were implemented using the basic tools of Python programming language, and the trials were run on a PC Intel I5 8265U.

The simulations conduct a search over the gridded domain of the size 80×80 cells. In the illustrations below, we consider the group of three agent,s searching for three static targets. Each agent is equipped with sensors of two types.

Starting positions of the searchers are (5, 5), (8, 8), and (62, 62), and locations of the targets are (65, 76), (75, 70), and (75, 78).

To obtain the lower bound of search time, we consider the scenario in which all the search agents have complete information about the targets' locations and move directly toward the targets. In this

Table 3.1 Search time in different scenarios with respect to the frequency of false alarms.

False alarms per time unit	Search time			
	Lower bound	Global map	Partial data sharing	Individual agents' maps
800	158	166	185	225
1600	158	188	202	321
3200	158	229	274	361

case, the overall search time by three agents is $T_{\min} = 158$, and this is the minimal possible time of search in the considered setup. The other search scenarios were compared with this time $T_{\min}$.

In the first series of simulations, we considered the search with constant sensors' sensitivity $\lambda(\mathbb{s}) = 20$ and different ratios of false alarms. The threshold probability is $P_k^* = 0.75$ for each type k of the sensors. Results of the simulations are summarized Table 3.1.

As it was expected, the best results were obtained in the scenario in which all three searchers share complete information and then use global probability map created using this common knowledge. In this case, the time of search is greater than its lower bound only in 5% (for 800 false alarms per time unit), 19% (for 1600 false alarms per time unit), and 45% (for 3200 false alarms).

The search with partial information sharing obtains the worthier results. In this case, the time of search is greater than its lower bound in 17% (for 800 false alarms per time unit), 28% (for 1600 false alarms per time unit), and 73% (for 3200 false alarms).

Finally, the worst results were obtained in the search where the agents do not share information and use their individual probability maps. In this case, the time of search is greater than its lower bound in 42% (for 800 false alarms per time unit), 103% (for 1600 false alarms per time unit), and 128% (for 3200 false alarms).

Increase of the search time with the frequency of false alarms observed in all scenarios represents the agents' reaction to the greater uncertainty in the data about the targets' locations.

In the next series of simulations with the same agents, we considered the dependence of search time on the threshold probability P_k^* in the scenarios with partial information sharing between the search agents. The obtained dependence is shown in Figure 3.3.

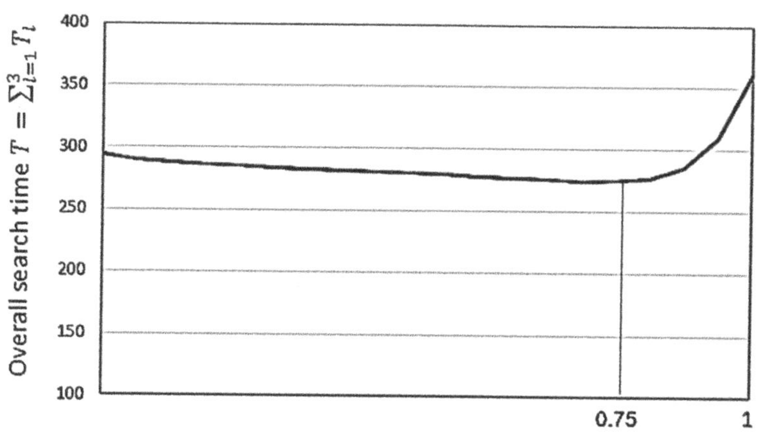

Figure 3.3 Dependence of the search time T on the threshold probability P_k^*.

Table 3.2 Search time in different scenarios with respect to the sensors' sensitivity.

Sensors' sensitivity	Search time			
	Lower bound	Global map	Partial data sharing	Individual agents' maps
const $\lambda = 20$	158	229	274	361
$\alpha = 1.05$	158	195	209	280
$\alpha = 1.10$	158	191	195	220
$\alpha = 1.20$	158	178	182	200

Minimal time is reached for the threshold probability $P_k^* = 0.75$. Note that for the values $P_k^* > 0.75$, the search time increases exponentially, while for the probabilities $P_k^* < 0.75$ this time slowly decreases and its decrease is close to linear.

These simulations show that the value of the threshold probability P_k^* is crucial in the cooperative search by the group of the agents and can completely change search results.

In the last series of simulations, we considered dependence of the search time on the sensitivity of the sensors. The obtained results are summarized in Table 3.2.

In all scenarios, the frequency of the false alarms is 3200 per time unit.

The obtained results support the expectation that greater sensitivity results in a shorter search time and demonstrate the effectiveness of dynamic sensitivity tuning.

3.4 Conclusions

In the chapter, we considered a basic procedure of search by a group of search agent and formulated the problems, which appear in such search. In the procedure, we assumed that the detection can include both false-positive and false-negative errors.

For both types of errors, we considered three levels of communication and information sharing:

- complete information sharing in which the probability map of each searcher is available to all other searchers;
- partial information sharing in which the searchers share the most robust parts of their probability maps;
- no information sharing in which each search agent uses its individual probability map and has no access to the probability maps of the other searchers.

In these scenarios, we also considered constant and varying sensitivity of the sensors.

For the considered scenarios, we defined the methods of decision-making, sensor fusion, and information sharing. These methods on one hand represent the data and control flows in the system and include the processing of false-positive and false-negative detection errors, and, on the other hand, are simple enough for practical applications.

The developed models were implemented in the Python software. Simulations show that, as it was expected, the shortest search times were demonstrated by the groups of agents with complete information sharing and the longest search times by the groups without information sharing. Partial information sharing results in the intermediate time searches.

Online tuning of the sensors' sensitivity allows for shortening the search times in all three considered cases of information sharing; however, the influence of the levels of information sharing is still the same.

4

Reactive Detection Algorithms

In this chapter, we define behavior of the group of search agents moving in a gridded domain with the aim of detection of hidden targets. It is assumed that the agents observe the environment by their on-board sensors and can recognise the targets from certain nonzero distances.

4.1 Agents' Policies and Decision-making

Let $C = \{c_1, c_2, \ldots, c_n\}$ be a finite set of cells. We assume that there is a set of m, $1 \le m \ll n$, search agents searching for hidden targets in the set C.

Denote by $\tau_j(T) = (c_j(0), c_j(1), \ldots, c_j(T))$ the trajectory of jth agent, $j = 1, 2, \ldots, m$, which starts in the cell $c_j(0)$ and moves to the cell $c_j(T)$. A cell occupied by the search agents at time t is denoted by $c_j(T)$.

Being located at cell $c_j(t)$, the agent makes a decision about its next location $c_j(t+1)$ following a certain policy $\pi_j(P)$ that prescribes how to choose the next cell given a probability map P. As above, we assume that policy $\pi_j(P)$ for each jth agent does not depend on the time and for any t is specified by the applied probability map P.

The result of the application of the policy $\pi_j(P)$ is an action $\mathbb{a}_j(t)$, which defines the agent's move from the current cell $c_j(t)$ to the next cell $c_j(t+1)$. More precisely, the policy is a function $\pi_j : P \to \mathbb{A}$, where $\mathbb{A}$ is a set of possible actions, and an action is defined as a function $\mathbb{a} : C \to C$ that specifies the choice of the agent's positions.

Assume that there are ξ targets, $\xi < n$, distributed somewhat over the domain, and recall that $p^{\text{global}}_{s_i=1}(t)$ is the probability of detecting the targets in cells c_i, $i = 1, 2, \ldots, n$, by the group of agents. We assume that both the number ξ of the targets and the global probabilities $p^{\text{global}}_{s_i=1}(t)$ are unknown to the agents.

Denote by $T_\theta(p \mid \pi_1(P), \pi_2(P), \ldots, \pi_m(P))$ the time required to detect the target θ, $\theta = 1, 2, \ldots, \xi$, with probability p given the agents' policies $\pi_1(P), \pi_2(P), \ldots, \pi_m(P)$.

Recall (see Section 3.1) that following these assumptions, the goal is to define such policies that result in minimal time for detecting the last target

$$\left(\pi_1^*(P), \pi_2^*(P), \ldots, \pi_m^*(P)\right) = \underset{(\pi_1(P), \pi_2(P), \ldots, \pi_m(P))}{\text{argmin}} \max_{\theta=1,2,\ldots,\xi} T_\theta(p).$$

As it was noted, this problem is computationally hard and cannot be solved directly by conventional optimization techniques.

To approximate the policies $\pi_1^*(P), \pi_2^*(P), \ldots, \pi_m^*(P)$, we follow three different approaches:

- maximization of the expected information gain (EIG) over the cells that are reachable by each of the agents in a single move;
- heading the agents toward the center of view (COV), which is the cell that provides maximum EIG over all the cells in the domain;
- heading the agents toward the center of the distribution – the center of gravity (COG), which is defined by the first moment of the probability map.

Note that the last approach is a greedy heuristic that requires minimal computation efforts, while the first two approaches are more complicated heuristics that require computation of the possibilities for local or global neighborhoods of each agent.

Let $P^{\text{sensor}}(j, k, t \mid \mathbb{a})$ be a sensor probability map $P^{\text{sensor}}(j, k, t \mid \mathbb{a})$ as obtained after executing an action $\mathbb{a}$ by the agent j and $P^{\text{sensor}}(j, k, t \mid \odot)$ be a sensor probability map obtained without movement, where $\odot$ denotes the action "stay in the current position."

The value EIG(j, k, t) of EIG for each sensor $\mathbb{s}_{jk}$ of the jth agent at time t is defined by the sum of the Kullback–Leibler distances between these two maps:

$$\text{EIG}(j, k, t) = D_{KL}(P^{\text{sensor}}(j, k, t \mid \mathbb{a}_j) \| P^{\text{sensor}}(j, k, t \mid \odot)),$$

where

$$D_{\text{KL}}(p \| q) = \sum_x p(x) \log(p(x)/q(x))$$

and logarithm is taken to the base of 2.

Certainly, instead of the Kullback–Leibler distance that is a pseudo-metric in the probability distributions space, other information-theoretic metrics can be used. In particular, EIG can be defined by Jensen–Shannon divergence

$$\frac{1}{2} D_{\text{KL}}(p \| M) + \frac{1}{2} D_{\text{KL}}(q \| M), M = \frac{1}{2}(p + q),$$

or using Ornstein or Rokhlin metrics; for application of such metrics to search problems, see [35]. Nevertheless, here we use the conventional definition of the EIG, and since the heuristics do not need the metric properties of the probability distributions space, we do not require the distance function to be a formal metric.

At the level of the jth search agent, the EIG(j, t) is defined as a sum of the EIGs for each sensor $\mathbb{s}_{jk}$,

$$\text{EIG}(j, t) = \sum_{k=1}^{l} \text{EIG}(j, k, t).$$

Similarly, for the group of m agents, the EIG(t) at time t is a sum of the agents' gains

$$\text{EIG}(t) = \sum_{j=1}^{m} \text{EIG}(j, t).$$

Note that instead of calculating Kulback–Leibler distances for each agent based on his own map and calculating global probability maps over all agents, the EIGs of higher levels are calculated by the sums of EIGs of the lower levels. Such definition follows a line of additive information property leading to simpler computations.

Using the EIGs, the agents choose the next actions such that they maximize the EIG value

$$\mathbb{a}^*(t) = \underset{\mathbb{A}}{\text{argmax}} \ \text{EIG}(t).$$

For example, being in any cell c_i an agent can choose one of four possibilities (see Section 2.1)

$$\mathbb{a}(t) \in \mathbb{A} = \{\uparrow, \rightarrow, \downarrow, \leftarrow, \odot\}$$

or one of nine movement possibilities

$$\mathbb{a}(t) \in \mathbb{A} = \{\uparrow, \nearrow, \rightarrow, \searrow, \downarrow, \swarrow, \leftarrow, \nwarrow, \odot\}$$

where, as in Section 2.1, arrows define the directions of the steps, and $\odot$ means "to stay in the current position."

The sensor probabilities $p^{\text{sensor}}_{s_i=1}(j, k, t \mid \mathbb{a})$ given the agent action $\mathbb{a}$ can be defined either by using the global probability map, $P^{\text{global}}(t)$, or by using the agent probability map, $P^{\text{agent}}(t)$. In the former case, the sensor probabilities are:

$$p^{\text{sensor}}_{s_i=1}(j, k, t \mid \mathbb{a}) = p^{\text{global}}_{s_i=1}(t) Pr\{\tilde{a}(c_i, t) = 1 \mid s_i(t) = 1\} \exp[-d(c_i, c(\mathbb{a}))/\lambda_{jk}],$$

$$p^{\text{sensor}}_{s_i=1}(j, k, t \mid \odot) = p^{\text{global}}_{s_i=1}(t) \times Pr\{\tilde{a}(c_i, t) = 1 \mid s_i(t) = 1\} \exp[-d(c_i, c(\odot))/\lambda_{jk}],$$

and in the latter case these probabilities are

$$p^{\text{sensor}}_{s_i=1}(j, k, t \mid \mathbb{a}) = p^{\text{agent}}_{s_i=1}(j, t) Pr\{\tilde{a}(c_i, t) = 1 \mid s_i(t) = 1\} \exp[-d(c_i, c(\mathbb{a}))/\lambda_{jk}],$$

$$p^{\text{sensor}}_{s_i=1}(j, k, t \mid \odot) = p^{\text{agent}}_{s_i=1}(j, t) Pr\{\tilde{a}(c_i, t) = 1 \mid s_i(t) = 1\} \exp[-d(c_i, c(\odot))/\lambda_{jk}],$$

where $c(\mathbb{a})$ is the agent's location after processing the action $\mathbb{a}$, and $c(\odot)$ is the agent's location if it decides to stay in the current cell, $i = 1, 2, \ldots, n$.

The second approach to govern the agents over the set C uses the COV and aims toward the cell that provides maximum EIG over the cells in the set C. It means that in contrast to EIG calculated over neighboring locations, the COV approach to the EIG over all the cells in the domain is applied.

Thus, instead of calculating the sensor probabilities by equations using distances $d(c_i, c(\mathbb{a}))$ and $d(c_i, c(\odot))$, in the COV approach calculation, the sensor probabilities are defined by using the distances $d(c_i, c_\eta)$ between the cell c_i and other cells in the domain, c_η, $\eta = 1, 2, \ldots, n$, that can be considered as candidate locations of the COV.

In parallel to the previous equations, in this case for the cell c_i, $i = 1, 2, \ldots, n$, we have

$$p^{\text{sensor}}_{s_i=1,G}(j, k, t \mid c_\eta) = p^{\text{global}}_{s_i=1}(t) Pr\{\tilde{a}(c_i, t) = 1 \mid s_i(t) = 1\} \exp[-d(c_i, c_\eta)/\lambda_{jk}],$$

$$p^{\text{sensor}}_{s_i=1,A}(j, k, t \mid c_\eta) = p^{\text{agent}}_{s_i=1}(j, t) Pr\{\tilde{a}(c_i, t) = 1 \mid s_i(t) = 1\} \exp[-d(c_i, c_\eta)/\lambda_{jk}].$$

If the agent chooses to stay in its current location, then the distance is defined as above by $d(c_i, c(\odot))$ and the sensor probabilities $p^{\text{sensor}}_{s_i=1}(j, k, t \mid \odot)$ and $p^{\text{sensor}}_{s_i=1}(j, k, t \mid \odot)$ are calculated by the above given equations.

Using these sensor probabilities, the EIG_η for the cell $\eta = 1, 2, \ldots, n$, is defined as follows:

$$\text{EIG}_\eta(j, k, t) = D_{\text{KL}}(P^{\text{sensor}}(j, k, t \mid c_\eta) \| P^{\text{sensor}}(j, k, t \mid \odot)),$$

$$\text{EIG}_\eta(j, t) = \sum_{k=1}^{l} \text{EIG}_\eta(j, k, t),$$

$$\text{EIG}_\eta(t) = \sum_{j=1}^{m} \text{EIG}_\eta(j, t),$$

and the COV is the cell in which EIG_η reaches its maximum

$$\text{COV}(t) = \underset{c_\eta}{\text{argmax}}\ \text{EIG}_\eta(t).$$

Finally, the COG, which is the first moment of the probability map, is calculated as follows. In a two-dimensional domain, location of each cell c_i, $i = 1, 2, \ldots, n$, is defined by two coordinates $c_i = (x_i, y_i)$. In addition, recall that $s_i = s(c_i) \in \{0, 1\}$ stands for the state of the cell c_i. Then, the coordinates of the COG for the axes are

$$\mathrm{COG}_x(t) = \sum_{i=1}^{n} x_i p_{s_i=1}^{\mathrm{global}}(t) \Big/ \sum_{i=1}^{n} p_{s_i=1}^{\mathrm{global}}(t),$$

$$\mathrm{COG}_y(t) = \sum_{i=1}^{n} y_i p_{s_i=1}^{\mathrm{global}}(t) \Big/ \sum_{i=1}^{n} p_{s_i=1}^{\mathrm{global}}(t),$$

and the final location of the COG is obtained by rounding the values $\mathrm{COG}_x(t)$ and $\mathrm{COG}_y(t)$ to the closest integers that is

$$\mathrm{COG}(t) = ([\mathrm{COG}_x(t)], [\mathrm{COG}_y(t)]).$$

Note that since we consider only the states $s_i = 1$, the sum of the probabilities in the denominator differs from the unit and varies with time and with the number of targets.

Since both here and in the previous case the desired cells COG(t) and COV(t) can be located far from the current agent's location, the agent follows toward these cells by steps and changes its direction with respect to the changes of the coordinates of both COG(t) and COV(t).

4.2 Policies Control and Brute-Force Learning

To control the obtained policies, we apply a simple look-backward method. Following this method the agents' policies are controlled as follows.

Let $P^{\mathrm{global}}(t-1)$ be a global probability map at time $t-1$ and assume that following the chosen policies $\pi_j(P^{\mathrm{global}}(t-1))$, $j = 1, 2, \ldots, m$, the agents made a decision and conducted corresponding actions. Then, at time t each of the agents is located in a new cell, and following observations from these cells a new global probability map $P^{\mathrm{global}}(t)$ is constructed. The value

$$V_\pi(t) = D_{\mathrm{KL}}\left(P^{\mathrm{global}}(t) \middle\| P^{\mathrm{global}}(t-1)\right),$$

is an actual information gain that was obtained by the actions defined by the policy $\pi = (\pi_1, \pi_2, \ldots, \pi_m)$.

Based on the policy π, at time t each agent makes a decision regarding its action. After the action and the corresponding movement, the agent observes the environment, and following the obtained sensor maps, the agents' probability maps and global probability map are refined. In parallel, to control the agent's policy, the value of information gain $V_\pi(t)$ is calculated. This process is illustrated in Figure 4.1.

The calculated information gain $V_\pi(t)$ indicates the efficiency of the applied policy, and it is used as a comparative measure of the agents' decisions.

The accumulated value of information gains up to some time T gets larger as the agents' policies are more efficient in the sense of the quantity of the obtained information about the targets' locations. Accordingly,

$$\pi^*(T) = \underset{\pi}{\mathrm{argmax}} \sum_{t=1}^{T} V_\pi(t)$$

is the best policy among all combinations of the agents' policies, π_j, $j = 1, 2, \ldots, m$, which can be applied to the available global probability maps util time T.

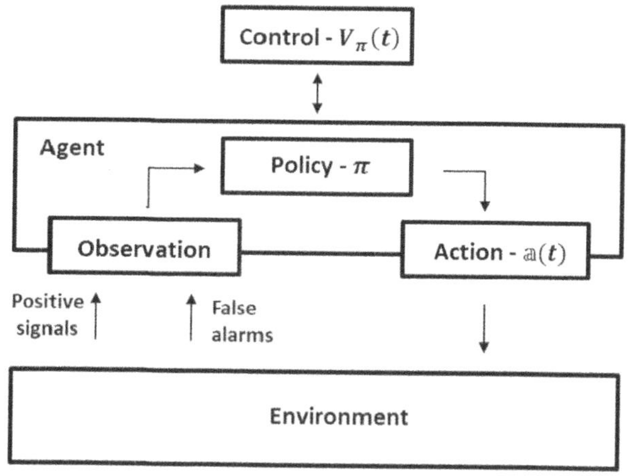

Figure 4.1 The process of the agents' decision-making and control.

Learning measure is also based on actual information gain, but it is defined over the actions $a = (a_1, a_2, \dots, a_m)$ that were chosen by the agents

$$V_a(t) = D_{KL}\left(P^{global}(t \mid a) \middle\| P^{global}(t \mid \odot)\right),$$

where $P^{global}(t \mid a)$ stands for the global probability map obtained after performing the actions of all the agents, and $P^{global}(t \mid \odot)$ is the global probability map in the situation where all the agents stay at their current locations.

Selection of the actions is conducted as follows. Assume that at time t the agents are at their locations and observe a certain global probability map $P^{global}(t)$. Then, for each combination $a = (a_1, a_2, \dots, a_m)$ of their actions and for the null action $\odot = (\odot_1, \odot_2, \dots, \odot_m)$, global probability maps $P^{global}(t \mid a)$ and $P^{global}(t \mid \odot)$ are obtained, and an average information gain $\langle V_a \rangle(t)$ for the combination a is specified. The best combination of actions is defined by the maximal value $\langle V_a \rangle(t)$ that is

$$a^*(t) = \underset{a}{\operatorname{argmax}} \langle V_a \rangle(t).$$

This is the brute-force learning that requires consideration of all possible combinations of actions for all the agents with large number of iterations. Thus, in practice, such a policy cannot be carried out for large number of agents and actions. However, for a single agent, this learning step can be performed in a short time with available computation resources.

Below the brute-force learning is used as a reference for evaluating the considered methods.

4.3 Numerical Simulations

As above, numerical simulations were implemented using the Python programming language, executed by a PC Intel I5 8265U processor. The search is conducted over a gridded square domain of size $n = n_1 \times n_2$ cells, and it is assumed that each agent and each target can occupy only one cell in the domain.

In all cases, unless defined specifically, the run times of the algorithms are indicated by the number of iterations. Also, in all tables the policies are measured with respect to the defined approaches: EIG, COV, and COG.

The simulated scenarios included different numbers, m, of the search agents and different numbers l of the targets. Also, we assumed that there are two types of sensors and each jth agent is equipped with two sensors $\mathbb{s}_{j1}$ and $\mathbb{s}_{j2}$ of different types with the sensitivities λ_{j1} and λ_{j2}, correspondingly. Both true and false alarms are sent with respect to the sensors' types and are perceived separately by each of the two sensor types.

Following the goal of finding such policies that result in a minimal time of detecting the last target, we defined maximal simulation time by

$$T_{\max}(p) = \max_{\theta=1,2,\dots,\xi} T_\theta(p),$$

where, as above, $T_\theta(p)$ is the time required to detect the target θ, $\theta = 1, 2, \dots, \xi$, with probability p. Then, the policies that result in minimum time $T_{\max}(p)$ given probability p are the best policies. In the simulations, we used the probability $p = 0.95$.

The results presented below were obtained by averaging the outcomes of five repeated trials, each containing 30 sessions executed with the same parameters, with the same initial locations of both the searchers and the targets. The true and false alarms in the sessions were generated by the same uniform distribution with a random seed for each trial.

4.3.1 Detection by a Single Agent

Let us start with a small illustrative example of detection by a single search agent. To simulate the brute-force learning (see Section 4.2), we considered a small domain of the size $n = 20 \times 20 = 400$ cells. A broadcast of the false alarms was distributed uniformly over the domain, and the frequency of sending false alarms from all 400 cells was 100 false alarms per time unit for each type of sensor, that is, on average 1/4 alarms per time unit from each cell to each type of sensor. The sensitivities of the sensors are $\lambda_{11} = \lambda_{12} = 10$.

In the first set, a single search agent was detecting $l = 3$ targets located in the cells with coordinates $c_1 = (11, 16)$, $c_2 = (0, 14)$, and $c_3 = (7, 1)$; starting position of the agent was $c(0) = (20, 8)$. Results of the simulation trials are summarized in Table 4.1.

As it was expected, the best result with minimal search time $T_{\max}(0.95)$ is reached by the brute-force learning, while the times of search following the policies based on the EIG, the COV, and the COG are close to this best result. Note that since detection is conducted by a single agent, the results obtained by the COV and COG policies are equal.

Activity of the single searcher agent detecting $l = 3$ targets using the COV policy is illustrated in Figure 4.2.

The values

$$V_\pi(T) = \sum_{t=1}^{T} V_\pi(t)$$

Table 4.1 Times required for the detection of the last among $l = 3$ targets with probability $p = 0.95$ by a single agent implementing different policies.

	Detection times			
Detection policy	**First target**	**Second target**	**Third target**	$T_{\max}(0.95)$
Static search agent	15	72	15	72
EIG	15	19	11	19
COV	8	18	15	18
COG	8	18	15	18
Brute-force learning	10	17	14	17

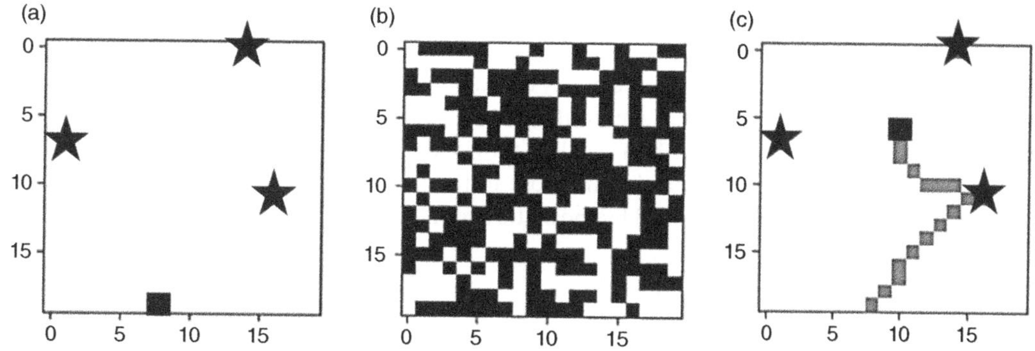

Figure 4.2 Activity of the single search agent detecting three targets: (a) targets' locations (stars) and initial agent's location (squire); (b) 100 false alarms per time unit for each type of sensor (white color indicates false alarms); (c) the agent's trajectory.

Table 4.2 Accumulated information gain in the detection of $l = 3$ targets by a single agent for the times $T = 10$ and $T = 17$.

	Accumulated information gain $V_\pi(T)$	
Detection policy	$T = 10$	$T = 17$
Static agent	3.6	4.1
EIG	4.2	6.1
COV	4.4	6.4
COG	4.4	6.4
Brute-force learning	4.6	6.8

of the accumulated information gain that characterizes the effectiveness of the policy for the simulated policies are presented in Table 4.2. In the table, time $T = 17$ is the minimal time of detecting the last target based on the brute-force policy.

With respect to the detection time, maximal accumulated information gain $V_\pi(T)$ is obtained by the brute-force learning policy, while the EIG, COV, and COG policies result in the accumulated information gains that are close to the maximum.

In the next simulation, a single agent detects $l = 5$ targets located in the cells of the following coordinates: $c_1 = (11, 16)$, $c_2 = (0, 14)$, $c_3 = (7, 1)$, $c_4 = (5, 3)$, and $c_5 = (11, 15)$; the starting position of the agent is $c(0) = (19, 8)$. Results of the simulation trials are summarized in Table 4.3.

Table 4.3 Times required to detect the last among $l = 5$ targets with the probability $p = 0.95$ by a single search agent acting according to different policies.

	Detection times					
Detection policy	**First target**	**Second target**	**Third target**	**Fourth target**	**Fifths target**	$T_{max}(0.95)$
Static agent	14	94	29	12	9	94
EIG	9	32	20	13	7	32
COV	14	28	22	10	8	28
COG	14	28	21	10	8	28
Brute-force learning	14	25	17	18	11	25

Similarly to the previous case, the best result is obtained by brute-force learning and the times of search following the policies based on the COV and the COG are close to this best result. Note that in this case with larger number of targets, the policy based on the EIG is worse than the other policies.

Relations between the values of the accumulated information gain are again similar to the case of the detection of three targets. The best result is provided by the brute-force learning policy, the results of the COV and the COG policies are close to the brute force, and the EIG policy is less effective. The worst results are obtained by the static agent.

4.3.2 Detection by Multiple Agents

Finally, let us consider the process of detection by a group of agents that use both global probability map, which represents the knowledge of the group, and share information and use probability maps of each other.

Since, in this case, we have not considered brute-force learning, in the simulations, we used a greater domain of size $n = 40 \times 40 = 1600$ cells. As mentioned above, a broadcast of the false alarms was distributed uniformly over the domain; however, the frequency of sending false alarms from all 1600 cells was 400 false alarms per time unit for each type of sensor that is 1/4 alarms per time unit on average from each cell to each type of sensor. It is assumed that each agent is equipped with sensors of two types; and the sensitivities of each type are denoted by λ_{j1} and λ_{j2}.

In the simulations we checked activity of the search agents that follow the EIG, COV, and COG policies and use different scenarios of decision-making:

- decisions based on the agents' probability maps *versus* decisions based on the global probability map;
- separate choice of action by each agent *versus* mutual selection of actions by all the agents in the group.

Consider the same setting as in the previous simulations, namely detection of $l = 5$ targets located at the cells with coordinates $c_1 = (4, 34)$, $c_2 = (6, 23)$, $c_3 = (37, 3)$, $c_4 = (32, 13)$, and $c_5 = (2, 5)$. The search is conducted by $m = 2$ search agents, each of which is equipped by two sensors with respective sensitivities $\lambda_{j1} = \lambda_{j2} = 10, j = 1, 2$. Initial locations of the agents are $c_1(0) = (25, 3)$ and $c_2(0) = (20, 9)$. Activity of these two agents using the EIG policy is illustrated in Figure 4.3.

Figure 4.3 Activity of two search agent detecting five targets: (a) targets' locations (stars) and initial agents' locations (squares); (b) 400 false alarms per time unit for each type of sensor (white color indicates false alarms); (c) the agent's trajectories.

Table 4.4 Times required for detecting the last among $l = 5$ targets with the probability $p = 0.95$ by $m = 2$ search agents acting according to different policies.

Detection policy		Detection times					
		First target	Second target	Third target	Fourth target	Fifth target	$T_{max}(0.95)$
Static agents		400	90	47	27	70	400
EIG	Agents' maps agents' actions	95	37	51	18	60	95
	Global map agents' actions	143	123	53	28	109	143
	Global map group actions	138	116	35	28	103	138
COV	Agent's maps agents' action	88	33	43	18	59	88
	Global map agents' actions	126	85	35	28	81	126
	Global map group action	108	85	37	32	61	108
COG	Agent's maps	87	33	43	18	59	87
	Global map	126	89	35	28	81	126

Results of the simulations are summarized in Table 4.4 (cf. Table 4.3).

The best results are provided by the policies based on the COV and the COG; these policies result in the same detection times. The EIG policy results in a longer search than the COV and the COG policies, yet, better than the static agent policy.

Note that the decision-making policies based on the agents' maps provide significantly better results than those based on the global map. In other words, more information is not always better in the detection tasks unless actions between the agents can be synchronized.

The reason for this result is the following. Relying on a single global map, all agents aim at the same preferable regions with a higher probability of detecting the targets and ignore the regions with lower probabilities. However, because of false-positive and false-negative errors, the targets can appear in those ignored regions, to which the agents return only after the unsuccessful detection in the preferable regions.

In contrast, while each search agent uses its own probability map the agent considers its local region and continues to the other regions only after unsuccessful detection in its close neighborhood. In such a manner, the agents divide the task and conduct detection process in parallel in different regions. The global map is used for terminating the detection process.

The accumulated information gain for the considered detection of $l = 5$ targets by $m = 2$ search agents is sumarized in Table 4.5.

The obtained results support the previous observations. The higher accumulated information gain is obtained by the COV and the COG policies, the EIG policy obtains the worse results, and the lowest gain is obtained by the group of static agents. Also, better results are achieved by decision-making policies based on the agent maps, and a better choice of actions is provided by using group action.

Table 4.5 Accumulated information gain in the detection of $l = 5$ targets by $m = 2$ agents for the times $T = 75$ and $T = 100$.

	Detection policy	Accumulated information gain $V_\pi(T)$	
		$T = 75$	$T = 100$
	Static agents	4.4	6.5
EIG	Agents' maps/agents' actions	9.9	13.6
	Global map/agents' actions	5.7	7.9
	Global map/group action	5.8	8.7
COV	Agents' maps/agents' actions	10.8	14.2
	Global map/agents' actions	8.4	11.7
	Global map/group action	9.2	12.6
COG	Agents' maps	10.9	14.2
	Global map	8.4	11.7

Table 4.6 CPU times (in seconds) required for detecting $l = 5$ targets by $m = 2$ agents with probability $p = 0.95$.

	Detection policy	CPU time (sec)
	Static agents	240
EIG	Agents' maps/agents' actions	75
	Global map/agents' actions	180
	Global map/group action	850
COV	Agents' maps/agents' actions	220
	Global map/agents' actions	350
	Global map/group action	3700
COG	Agents' maps	25
	Global map	32

Along with that the better information gain and search time are not necessarily correlated with the time required for detection of the targets. Table 4.6 summarizes the CPU times (in seconds for above-indicated PC) required to detect the targets using different policies.

To check the relation between detection efficiency and sensors' sensitivity, the same detection scenario was simulated for the agents that are equipped with sensors with different sensitivities $\lambda_{1k} = 12$ and $\lambda_{2k} = 8$, $k = 1, 2$. Results of these simulations are summarized in Table 4.7, which presents detection times $T_{\max}(0.95)$ and that accumulated information gains $V_\pi(100)$ for different search policies.

Change of the sensors' sensitivities preserves the already considered trends in the efficiencies of different search policies, but also stresses an advantage of the group action relative to the actions by each agent separately.

Table 4.7 Detection times and accumulated information gain in the detection of $l = 5$ targets by $m = 2$ agents with the sensor sensitivities $\lambda_{1k} = 12$ and $\lambda_{2k} = 8$, $k = 1, 2$.

	Detection policy	$T_{max}(p = 0.95)$	$V_\pi(T = 100)$
	Static agents	300	5.1
EIG	Agents' maps/agents' actions	81	13.6
	Global map/agents' actions	144	6.2
	Global map/group action	129	9.0
COV	Agents' maps/agents' actions	63	12.9
	Global map/agents' actions	109	11.7
	Global map/group action	96	12.4
COG	Agents' maps	67	12.9
	Global map	109	11.9

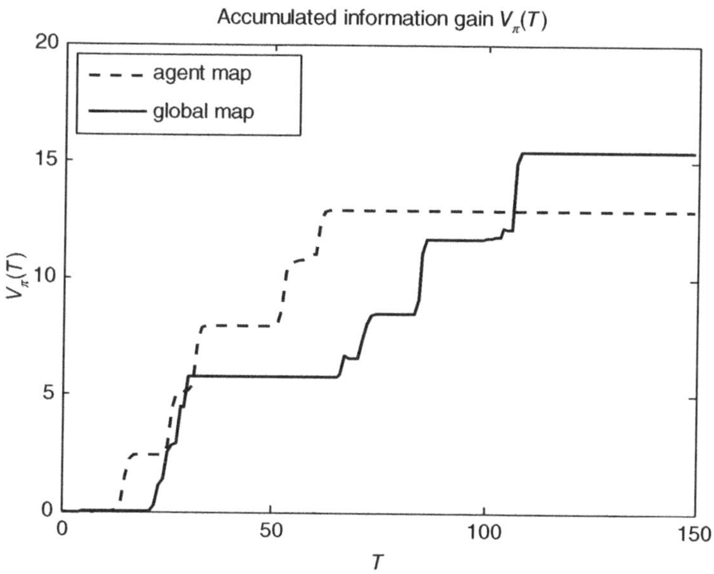

Figure 4.4 Dependence of the accumulated information gain $V_\pi(T)$ on the detection time T for the COV policy with agent action choice; $l = 5$ targets, $m = 2$ agents, and sensors' sensitivities $\lambda_{1k} = 12$ and $\lambda_{2k} = 8$, $k = 1, 2$.

Finally, let us consider dependence of the accumulated information gain on the detection time. An example of such dependences is presented in Figure 4.4, which shows the relation between the detection time of $l = 5$ targets by $m = 2$ agents with different sensors sensitivity $\lambda_{1k} = 12$ and $\lambda_{2k} = 8$, $k = 1, 2$, following the COV policy and the accumulated information gain.

At the beginning of the detection process, the accumulated information gain increases faster for the policy based on the global map. However, as the detection process continues, the accumulated information gain obtained by the global map policy converges to a value significantly greater than the accumulated information gain obtained by the policy based on the agents' maps.

Similar simulations demonstrated that same trends hold for different settings and showed that as the number of agents increases, the difference between the best COV policy and the nearly best COG and EIG policies also increases.

4.4 Conclusions

The chapter presents three heuristic techniques for navigation of autonomous agents searching for hidden targets in the presence of false-positive and false-negative errors. Two of these heuristics are based on the EIG calculated over local neighborhood of each agent (EIG policy) or relative to the COV policy, and the third heuristic uses the COG policy of the targets' location probabilities. To make decisions regarding the next movements, the agents use either their probability maps or a global probability map.

Simulations show that in short-term detection processes, the policies based on the agents' maps outperform the policies based on the global map, both while the agents' movements are not synchronized (individual decision-making) and while the agents' actions are definitely synchronized (collective decision-making in the group).

However, in the long-term detections, the accumulated information gain during the search process using the global map policy was significantly larger than the accumulated information gain during the search process using the agents' maps policy. Probably, the reason for such a result is the following: while using the global map, the agents calculate information gain also taking into account the nonrelevant information based on the false alarms, and while using the agents' maps, the influence of such alarms is lower and, consequently, the accumulated information gain is lower as well.

Detection using the EIG policy demonstrated lower efficiency (in the terms of detection time) than the COV and COG policies while using both the agents' and the global probability maps. The main reason for such a result is the following. On the one hand, the EIG policy does not always recognize what should be the next step since the differences in the EIG obtained by staying in the current cell and by moving to a neighboring cell are extremely small and cannot be used for a reasonable selection of the actions. On the other hand, to reveal the center of view, the COV policy requires the agent to check all the cells in the domain and, consequently, succeeds in finding a significant change in the EIG.

While using the sensors with equal sensitivities, the COV and COG policies result in close or even equal detection times. Therefore, since the COG policy has extremely lower computational complexity than the COV policy, it should be preferred when the agents are equipped with similar sensors. However, if the sensitivities of the sensors are different, then the COV policy is significantly better.

As expected, a decision-making heuristic that relies on group actions leads to better performance than the one which relies on single agents' actions, yet the first heuristic requires much greater computation efforts. To shorten the running time, the number of calculations can be decreased by using some probability threshold. At each step of the computation, the cells with probabilities lower than the threshold are ignored. Then, the number of calculations is reduced without significantly influencing the quality of the search results.

Finally, the results obtained using the presented techniques are close to those obtained by the brute-force learning method. Such comparison both validates the methods and demonstrates that for the detection over large domains, and where brute-force learning cannot be used due to intractable computation complexity, these methods might provide sub-optimal results by reasonable computation efforts in reasonable running time.

5

Search with Learning

In the previous chapters, we considered the methods of search and detection in which the searchers react to the changes of the probability maps and choose their steps following these changes. This chapter starts consideration of search algorithms with learning.

In this chapter, we recall the main concepts of reinforcement learning and stress the problems of supervised and unsupervised search (for detailed overview of these approaches, see e.g. [31, 38, 53]). In the next two chapters, these ideas will be used in the consideration of the search algorithms with deep Q-learning.

5.1 Reinforcement Learning

In the last decade, the use of reinforcement learning for navigating mobile agents was an objective of enormous number of works addressing different scenarios and tasks. For overview of the directions and results, see e.g. the papers [17, 38, 48]. Here we consider the basic ideas of these methods.

In Section 4.2, in the consideration of policy control, we already mention the brute-force learning that maximizes the Kullback–Leibler distances between probability maps.

As it is shown in Figure 4.1, at each moment t the agent being in a cell $c(t)$ observes the environment and following certain policy π chooses an action $\mathbb{a}(t)$. The quality of the action $\mathbb{a}(t)$ chosen in the cell $c(t)$ is characterized by the reward $R(c(t), \mathbb{a}(t))$.

By processing the chosen action $\mathbb{a}(t)$, the agent moves to the cell $c(t+1)$ (which may be equal to the cell $c(t)$ that means that the agent stays in the same cell) and the process continues with the time $t+1$.

The goal of the learning process is to define a policy π, which relates agent's location $c(t)$ with the action $\mathbb{a}(t)$ such that the total accumulated reward $\sum_t R(c(t), \mathbb{a}(t))$ reaches its maximum. In the deterministic algorithms, policy π is defined as a mapping that specifies the action with respect to the cell, and in the probabilistic case, policy π specifies the probabilities $\pi(\mathbb{a} \mid c)$ of choosing the action $\mathbb{a}$ given the searcher's cell c.

In the search for hidden targets, observation of the environment results in the knowledge of targets' location probabilities and their update with respect to the observation result. If the search agent recognises the target in a certain cell c, then it is assumed that the target is detected and probability $p(c)$ of target's location is decreased: in the case of erroneous detection, this probability is set to zero, otherwise to certain small value with respect to the value of detection error. The targets' location probabilities over the domain are updated respectively. The goal of the search agent is to detect the targets in minimal time or which is the same in minimal number of steps.

Since the search agent deals with the probabilities of the taregt's locations, the agent's policy is also defined in the terms of probabilities and the process is usually formulated as a Markov decision process (MDP) [64]. In general, the formulation is as follows.

Let $C = \{c_1, c_2, \dots, c_n\}$ be a finite set of cells, $\mathbb{A} = \{\mathbb{a}_1, \mathbb{a}_2, \dots, \mathbb{a}_l\}$ be a set of actions and $R : C \times \mathbb{A} \to \mathbb{R}$ be a reward function, where $\mathbb{R} = (-\infty, \infty)$ is a set of real numbers.

Denote by $\mathbb{P}(C)$ a set of probability distributions, $P(C)$, on the set C and by $\mathcal{P} : C \times \mathbb{A} \to \mathbb{P}(C)$ a state transition probability function that defines the probabilities $\mathcal{P}(c, \mathbb{a}, c')$ of moving from the cell c to the cell c' by action $\mathbb{a}$. Note that if the action $\mathbb{a}$ applied in the cell c unambiguously defines the next cell c', then probability $\mathcal{P}(c, \mathbb{a}, c')$ is equivalent to the probability of moving to the cell $c'(t+1)$ from the cell $c(t)$ at time t.

Then, maximal expected reward of the agent starting from the cell $c \in C$ is defined by the following Bellman equation [31]:

$$V^*(c) = \max_{\mathbb{a}} \left\{ R(c, \mathbb{a}) + \gamma \sum_{c' \in C} \mathcal{P}(c, \mathbb{a}, c') V^*(c') \right\},$$

where positive $\gamma < 1$ is a discount factor that provides convergence of the equations.

In the equation, the value $V^*(c)$ stands for the expected reward $V(c)$ obtained using optimal policy π^*. Consequently, given the function V^*, an optimal policy provides maximum of the expected reward

$$\pi^*(c) = \arg\max_{\mathbb{a}} \left\{ R(c, \mathbb{a}) + \gamma \sum_{c' \in C} \mathcal{P}(c, \mathbb{a}, c') V^*(c') \right\}.$$

In search and detection problems, the reward function $R\colon C \times \mathbb{A} \to \mathbb{R}$ and consequently the value function $V\colon C \times \mathbb{A} \to \mathbb{R}$ can be defined by different manners.

For example, if the goal is to obtain maximum information about the targets' locations, then the reward is defined in terms of expected information gain (see EIG value in Section 4.1). If the goal is to minimize the length of the searcher's path, then instead of maximization of the expected reward it is required to minimize an expected payoff, which is an expected length of the searcher's trajectory (see the use of COG and COV in Section 4.1). In Section 6.2.6, we formulate such minimization problem for the shortest path length (SPL) algorithm, and in Section 6.3.3 also consider the use of the center of gravity (COG) and center of view (COV), c.f. Section 3.3.1, where the one-step criterion is used.

Assume that the search agent implements the search using the EIG values. Then, the path of the agents is defined by the standard value iteration algorithm, which maximizes the expected reward (see e.g. 31).

The algorithm consists of two parts: calculation of the value function V^* and extraction of the optimal policy π^*. The algorithm is outlined as follows.

Algorithm 5.1 Path planning by value iteration

Input: Set $C = \{c_1, c_2, \dots, c_n\}$,
set $\mathbb{A} = \{\uparrow, \nearrow, \rightarrow, \searrow, \downarrow, \swarrow, \leftarrow, \nwarrow, \odot\}$ of possible movements,
reward function $R : C \times \mathbb{A} \to \mathbb{R}$,
transition probability function $\mathcal{P} : C \times \mathbb{A} \to \mathbb{P}(C)$,
discount factor $\gamma \in (0, 1)$,
termination threshold $\theta \in (0, 1)$.

Output: Optimal policy π.

<u>Calculating value function</u>

1. Create array $V = \{V_1, V_2, \dots, V_n\}$ with respect to the set C.
2. Create array $Q = \{Q_1, Q_2, \dots, Q_l\}$ with respect to the set $\mathbb{A}$.
3. Set $V_i = 0$, $i = 1, 2, \dots, n$.
4. Set $\delta = 0$.
5. Do:
6.   For each $i = 1, 2, \dots, n$ do:
7.   Set $V_0 = V_i$.
8.   For each $k = 1, 2, \dots, l$ do:
9.   Set $Q_k = R\left(c_i, \mathbb{a}_k\right) + \gamma \sum_{j=1}^{n} \mathcal{P}\left(c_i, \mathbb{a}_k, c_j\right) V_j$.
10.   End for.
11.   Set $V_i = \max_k (Q)$.
12.   Set $\delta = \max\{\theta, |V_i - V_0|\}$.
13.   End for.
14. While $\delta > \theta$.

<u>Extracting optimal policy</u>

15. Create array $\pi = \{\pi_1, \pi_2, \dots, \pi_n\}$ with respect to the set C.
16. Create array $A = \{A_1, A_2, \dots, A_l\}$ with respect to the set $\mathbb{A}$.
17. For each $i = 1, 2, \dots, n$ do:
18.   For each $k = 1, 2, \dots, l$ do:
19.   Set $A_k = R\left(c_i, \mathbb{a}_k\right) + \gamma \sum_{j=1}^{n} \mathcal{P}\left(c_i, \mathbb{a}_k, c_j\right) V_j$.
20.   End for.
21.   Set $k^* = \text{ind} \max_k (A)$ index of maximal element of the array A.
22.   Set $\pi_i = A_{k^*}$.
23. End for.
24. Return π.

The presented algorithm follows standard value iteration procedure; for the programming purposes, it can be written in shorter form.

The algorithm runs offline and creates a path for a single agent searching for static targets.

Now let us extend the value iteration algorithm to search by multiple agents.

A straightforward extension of the algorithm is obvious. Given a number m of search agents, state space is defined by the Cartesian product $\mathcal{C} = \underbrace{C \times C \times \dots \times C}_{m \text{ times}}$ and actions' set is defined by the Cartesian product $\mathcal{A} = \underbrace{\mathbb{A} \times \mathbb{A} \times \dots \times \mathbb{A}}_{m \text{ times}}$. Then, accordingly the reward function $\mathcal{R} : \mathcal{C} \times \mathcal{A} \rightarrow \mathbb{R}$, value function $\mathcal{V} : \mathcal{C} \times \mathcal{A} \rightarrow \mathbb{R}$, and transition probabilities $\mathcal{P} : \mathcal{C} \times \mathcal{A} \rightarrow \mathbb{P}(\mathcal{C})$ are defined.

However, such extension leads to insufficient computation complexity of the algorithm and requires a knowledge of the transition probabilities for a group of m agents, which is practically impossible. Along with that, for a given distribution of the targets' location probabilities available to each agent, the analysis can be essentially simplified using the maximum likelihood inference [26].

To overcome these problems, several approaches were developed. For example, Albrecht et al. [2] considered multi-agent reinforcement learning as a kind of game in which the agents can cooperate (cooperative games) or compete (antagonistic games). Bertsekas [3] suggested an approach where the agents share the reward and act sequentially maximizing this common reward. A detailed review of the methods of reinforcement learning in the multi-agent exploration tasks is presented in a paper by Garaffa et al. 17; for coordinated search see 46. Here we present an algorithm based on the Bertsekas approach 3 that inspired the methods presented below.

Let $C = \{c_1, c_2, \dots, c_n\}$ be a set of cells and let there be m search agents acting in C. Each kth agent, $k = 1, 2, \dots, m$, has a set $\mathbb{A}_k = \{\mathbb{a}_{k1}, \mathbb{a}_{k2}, \dots, \mathbb{a}_{kl}\}$ of possible movements and a reward function $R_k\colon C \times \mathbb{A}_k \to \mathbb{R}$.

Following formulation given in Chapter 3, assume that each kth agent manages its probability map $P_k = \{p_{k1}, p_{k2}, \dots, p_{kn}\}$, which includes the probabilities of targets' locations in the cells $c_i \in C$, $i = 1, 2, \dots, n$, from the point of view of the kth agent, $k = 1, 2, \dots, m$.

Assume that the kth search agent is the cell c. Being in this cell, the agent screens the cells reachable by the movements $\mathbb{a}_k \in \mathbb{A}_k$ and chooses a movement with respect to the probability $p_k \in P_k$ of the targets' location in the cell to which the movements draw (ties are broken randomly). For example, assume that the agent is in the c_i and by the action $\mathbb{a}_{k\eta}$ it moves to the cell c_j. Then, transition probability $\mathcal{P}(c_i, \mathbb{a}_{k\eta}, c_j) = p_{kj}$.

Path planning by value iteration for a group of m search agents with sequential decision-making is outlined as follows.

Algorithm 5.2 Multi-agent path planning by value iteration with sequential decision-making

---

Input: Set $C = \{c_1, c_2, \dots, c_n\}$,
number m of agents,
sets $\mathbb{A}_k$, $k = 1, 2, \dots, m$ of possible movements,
reward functions $R_k : C \times \mathbb{A}_k \to \mathbb{R}$, $k = 1, 2, \dots, m$,
initial probability maps P_k, $k = 1, 2, \dots, m$,
discount factor $\gamma \in (0, 1)$,
termination threshold $\theta \in (0, 1)$.

Output: Optimal policies π_k, $k = 1, 2, \dots, m$.

Calculating value function

1. Create empty probability map P_0.
2. Create arrays $V_k = \{V_{k1}, V_{k2}, \dots, V_{kn}\}$ with respect to the set C, $k = 1, 2, \dots, m$.
3. Create arrays $Q_k = \{Q_{k1}, Q_{k2}, \dots, Q_{kl}\}$ with respect to the sets $\mathbb{A}_k$, $k = 1, 2, \dots, m$.
4. Set $V_{ki} = 0$, $i = 1, 2, \dots, n$, $k = 1, 2, \dots, m$.
5. Set $\delta = 0$.
6. Do:
7.   For each $k = 1, 2, \dots, m$ do:
8.   Set $P_0 = \text{fuse}(P_0, P_k)$.
9.   For each $i = 1, 2, \dots, n$ do:
10.   Set $V_0 = V_{ki}$.
11.   For each $\eta = 1, 2, \dots, l$ do:
12.   Set $Q_{k\eta} = R_k\left(c_i, \mathbb{a}_{k\eta}\right) + \gamma \sum_{j=1}^{n} \mathcal{P}_0\left(c_i, \mathbb{a}_{k\eta}, c_j\right) V_{kj}$.
13.   End for.
14.   Set $V_{ki} = \max_k \left(Q_k\right)$.
15.   Set $\delta_k = \max\{\theta, |V_{ki} - V_0|\}$.
16.   End for.
17.   Set $\delta = \max\{\delta, \delta_k\}$.

18. End for.
19. Set
20. While $\delta > \theta$.

Extracting optimal policy

21. Create arrays $\pi_k = \{\pi_{k1}, \pi_{k2}, \dots, \pi_{kn}\}$ with respect to the set C, $k = 1, 2, \dots, m$..
22. Create arrays $A_k = \{A_{k1}, A_{k2}, \dots, A_{kl}\}$ with respect to the sets $\mathbb{A}_k$, $k = 1, 2, \dots, m$.
23. For each $i = 1, 2, \dots, n$ do:
24.   For each $k = 1, 2, \dots, l$ do:
25.   For each $\eta = 1, 2, \dots, l$ do:
26.   Set $A_{k\eta} = R_k\left(c_i, \mathbb{a}_{k\eta}\right) + \gamma \sum_{j=1}^{n} P_0\left(c_i, \mathbb{a}_{k\eta}, c_j\right) V_{kj}$.
27.   End for.
28.   Set $k^* = \text{ind}\max_{\eta}\left(A_k\right)$ index of maximal element of the array A_k.
29.   Set $\pi_{ki} = A_{k^*}$.
30.   End for.
31.   Set $k^* = \text{ind}\max_{k}(A)$ index of maximal element of the array A.
32.   Set $\pi_i = A_{k^*}$.
33. End for.
34. Return π_k, $k = 1, 2, \dots, m$.

In brief, this algorithm acts as follows: the first agent chooses the movement according to its probability map P_1, the second agent chooses the movement according to the fusion of the maps P_1 and P_2, the third according to the fusion of the maps P_1, P_2 and P_3 and so on until the last agent, which acts based on the global probability map. Without learning, this idea is implemented in Algorithm 2.1 and in the advanced form with sequential deep Q-learning in Algorithm 7.2.

Certainly, this algorithm is not even one, and there is a long series of algorithms of multi-agent reinforcement learning, which can be used in search and detection tasks. Overviews and taxonomy of multi-agent deep reinforcement learning methods appear in the papers by Nguyen et al. [44] and by Zhu et al. [67].

5.2 Search with Supervised and Unsupervised Learning

Activity of the systems with supervised learning follows two main stages:

1) learning stage during which the systems learn to map possible inputs to the desired outputs using a given data, and
2) acting stage while the systems predict the states of the environment using the learned map and act with respect to these predictions.

In some cases, the maps of the inputs to the outputs are called models, and the second stage is called prediction stage.

In the framework of the search and detection problems, supervised learning is implemented as follows. At the first stage the search agents are trained using certain scenarios of search, and at the second stage the search is conducted by the trained agents. For example, in search and rescue tasks [42], search agents are trained using a database of previous searches or using a set of generated search setups and simulated actions of the agents. During the search, the agents follow the trained trajectories and use results of the search for online training.

Learning methods implemented by the search agents depend on the considered situation and the agents' abilities. For example, in the search by a team of the agents, the process can start with clustering of the domain and continue with the use of regression models of decision trees by each agent acting in its cluster. In other situations, the agents can use different types of artificial neural networks that define the next steps of the agents given location probabilities of the targets.

In general, the process of search with supervised learning is outlined as follows.

Algorithm 5.3 Search with supervised learning

Input: Search domain,
motion and sensing abilities of the search agent,
number N of train sessions,
search time T or accuracy ε of search.

Output: Locations of the targets.

<u>Learning stage</u>

1. Start with maximal search time and zero detection probability.
2. Create initial search plan, e.g. random search.
3. Repeat N times:
4. Generate probability map.
5. Place the targets randomly with respect to the map.
6. Create search plan given previous plan and its evaluations:
   - the goal is minimal time with maximum detection probability.
7. Simulate search following the created search plan.
8. Evaluate search time and detection probability.
9. End repeat.

<u>Acting stage</u>

10. Observe probability map.
11. Create search plan.
12. While search time T is not expired or accuracy ε is not reached do:
13. Observe probability map.
14. Create search plan given previous search plan and its evaluations:
    - the goal is minimal time with maximum detection probability.
15. Conduct search following the created search plan.
16. Evaluate search time and detection probability.
17. End while.
18. Return locations of the targets.

Following this algorithm, the search agent trains on the simulated search plans and then searches the targets and trains during the search. If a database of previous real-world searches and their evaluations is available, then real search plans from the database are used during the learning stage instead of generated probability map and simulated searches.

Learning steps in Algorithm 5.1 are implemented in the lines 6 and 14 where current search plan is created on the basis of previous plan and results of its evaluations.

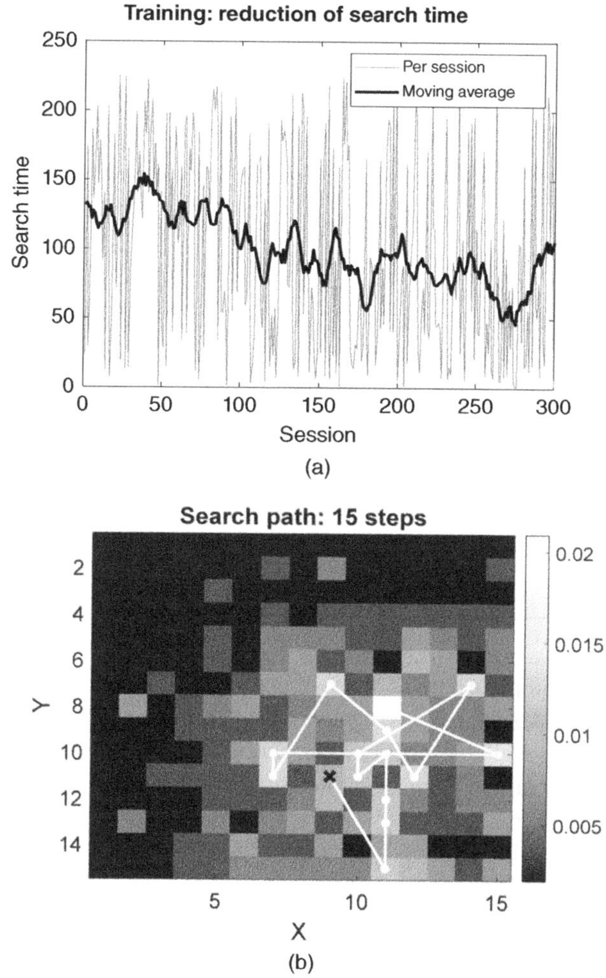

Figure 5.1 Actions of Algorithm 5.1: (a) reduction of average search time during training and (b) trajectory of the searcher (white line). Domain size $n = 15 \times 15$ cells, number of training sessions $N = 300$.

Figure 5.1 illustrates the actions of Algorithm 5.1 in the search for static target over the domain $n = 15 \times 15$ cells with $N = 300$ training sessions.

In the illustrated simulation, the searcher learns the probability map and then moves to the cells with maximal target location probabilities in the learned map.

The considered technique of supervised learning is useful in the search for static targets where probability map plays a role of the agent's memory. Following Bayesian updating, targets' location probabilities in the observed cells are decreased that avoids their further consideration, and targets' location probabilities in the remaining cells are increased that leads the searchers to consider these cells in further steps.

Along with that in the search for moving targets, supervised learning is less applicable, and in this case the methods of unsupervised learning are used. The task is not so much to learn a probability map and to train the searcher to follow this map, but to learn the patterns of the targets' movement and to predict their future locations.

Formally general algorithm of search with unsupervised learning literally follows the acting stage of the Algorithm 5.3. For completeness, we outline these stages below.

Algorithm 5.4 Search with unsupervised learning

Input: Search domain,
motion and sensing abilities of the search agent,
search time T or accuracy ε of search.

Output: Locations of the targets.

1. While search time T is not expired or accuracy ε is not reached do:
2. Observe probability map.
3. Create search plan given previous search plan and its evaluations:
   - the goal is minimal time with maximum detection probability.
4. Conduct search following the created search plan.
5. Evaluate search time and detection probability.
6. End while.
7. Return locations of the targets.

In this algorithm, search plan can include a complete path of the searcher or several steps toward the expected targets' locations. Figure 5.2 illustrates the actions of Algorithm 5.2: Figure 5.2a shows the search for three static targets without any available information about their locations, and Figure 5.2b shows the search for three randomly moving targets, while the probability map is available and updated with respect to the search results.

If the search agent has no information about locations of the targets, it sequentially screens the domain up to finding all the targets. Update of the probability map avoid returns to the already observed cell and since the targets are static, the zero probabilities remain during the next search.

If the search agent acts according to actual probability map, it moves toward the cells with maximal expected probability of targets' locations and updates the probabilities of already observed cells. However, since the targets are moving, the updated probabilities change with respect to the targets' motion, and the search agent can return to the already observed cells.

The considered algorithms are direct extensions of the reactive search algorithms in Chapter 2. In Chapters 6 and 7, we will see algorithms that follow the outline of Algorithm 5.3 and implement deep Q-learning techniques.

5.3 Two-Actor Search

As indicated above, in search and detection tasks unsupervised learning is mainly used for recognizing the patterns of the targets' movement. However, in certain scenarios such patterns can be predicted a priori or can be included in the search planning process.

Below, we consider two such scenarios: pursuit-evasion games in which the target escapes from being captured and actor-critic algorithm in which reinforcement learning includes an artificial critic that defines the Q-value of the actor's strategy.

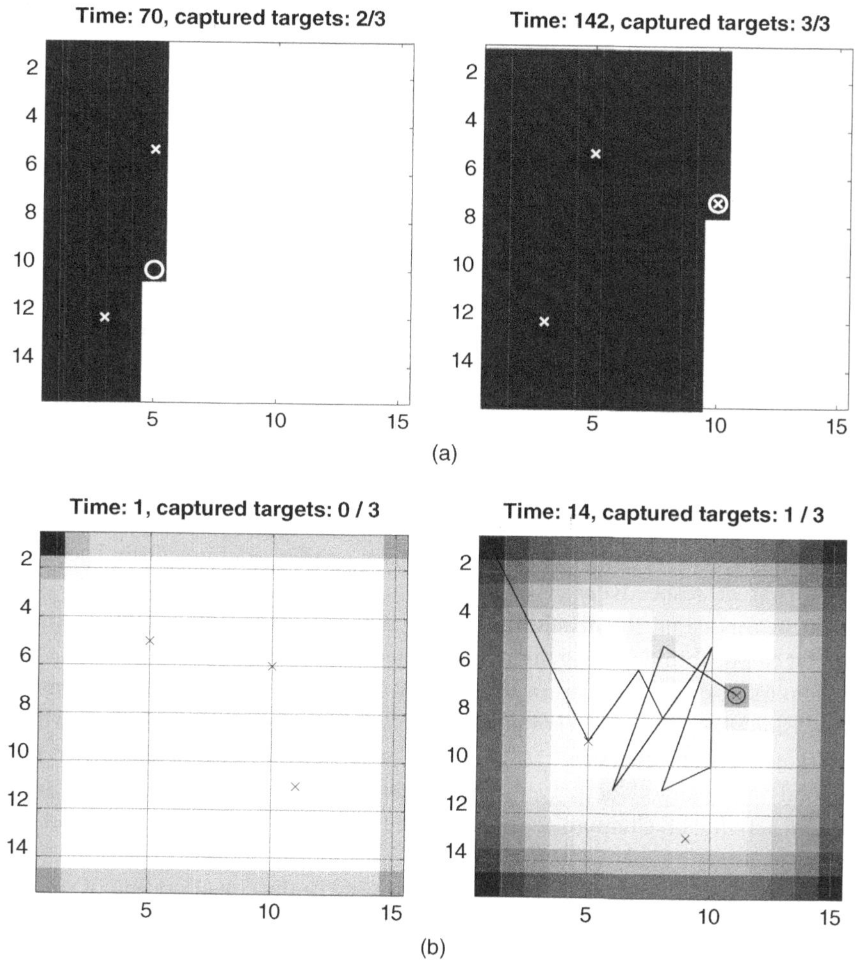

Figure 5.2 Activity of Algorithm 5.2: (a) search for three static targets and (b) search for three randomly moving targets (trajectory of the searcher is depicted by black line). Domain size is $n = 15 \times 15$ cells.

5.3.1 Pursuit-Evasion Game

Search games address the situations in which the target is aware of the searcher's actions and can try to escape from the searcher. In its turn, since the search agent can expect the target's reactions on its actions, it also can predict the target's behavior and can consider it while planning the search [1, 16] (for overview, see the book 35).

For illustration, let us consider a pursuit-evasion game with a single searcher and a single target. This game is a probabilistic version of "the Princess and the Monster" game 28 on a gridded domain. The game is formulated as follows 11.

Let, as above, $C = \{c_1, c_2, \ldots, c_n\}$ be a set of cells. Denote by $x(t) \in C$ the cell occupied by the target and by $y(t) \in C$ the cell occupied by the searcher at time t, $t = 0, 1, 2, \ldots, T$.

Assume that at any time t, searcher's observed area includes a single cell that is the searcher's location, that is $O(t) = \{y(t)\}$. In addition, assume that the detection is errorless, that is

$$z(x(t), O(t)) = \begin{cases} 1 & \text{if}\, x(t) \in O(t), \\ 0 & \text{otherwise.} \end{cases}$$

Then the probability of detecting the target in the cell $y(t)$ is equivalent to the probability $p_{\text{tar}}(y(t)) = Pr\{x(t) = y(t)\}$ that at time t the target is in the cell $y(t)$ occupied by the searcher.

Denote by $p_{\text{tar}}(c_i, t) = \{x(t) = c_i\}$ the probability that at time t the target is in the cell c_i and by $p_{\text{ser}}(c_i, t) = \{y(t) = c_i\}$ the probability that at time t the searcher is in the cell c_i. Then the expected number of rendezvous of the searcher and the target up to time T is [11]

$$\mathcal{E}(p_{\text{tar}}, p_{\text{ser}}, T) = \sum_{t=0}^{T}\sum_{i=1}^{n} z(c_i, \{c_j\})p_{\text{tar}}(c_i, t)p_{\text{ser}}(c_i, t).$$

The goal of the searcher is to find such function p_{ser}, which results in maximal expected number of the rendezvous with the target, and the goal of the target is to find such function p_{tar}, which results in minimal number of the rendezvous with the searcher.

The pair of the functions $\left(p^*_{\text{tar}}, p^*_{\text{ser}}\right)$ such that

$$\mathcal{E}\left(p^*_{\text{tar}}, p^*_{\text{ser}}, T\right) = \min_{p_{\text{tar}}} \mathcal{E}(p_{\text{tar}}, p_{\text{ser}}, T) = \max_{p_{\text{ser}}} \mathcal{E}(p_{\text{tar}}, p_{\text{ser}}, T)$$

is a solution of the game in the mixed strategies. The expected number $\mathcal{E}\left(p^*_{\text{tar}}, p^*_{\text{ser}}, T\right)$ of rendezvous is called the value of the search-evasion game, and the functions p^*_{tar} and p^*_{ser} define optimal policies of the searcher and of the target, respectively.

Existence of solution in this game in mixed strategies is guaranteed by the von Neumann's min-max theorem, and it can be found using general game-theoretic theorems (see e.g. [18]).

Usually, probabilistic pursuit-evasion games in the gridded domain are formulated in terms of competitive Markov decision processes or Markov games [14]. Solution of such games follows a line of optimization of Markov decision processes [24, 25, 64] in which transition probabilities in the searcher's motion depend on the target's activity and transition probabilities of the target's motion depend on the searcher's activity; cf. Chapters 2 and 3 where target's motion does not depend on the searcher's actions.

Game-theoretic models form a basis for different algorithms used in the consideration of search and detection problems from reinforcement learning [53] to the actor-critic algorithms [19]. Below, we briefly consider the actor-critic approach.

5.3.2 Actor-Critic Approach

The actor-critic algorithms extend the reinforcement learning techniques (see Section 5.1).

Recall that in the reinforcement learning for each state c two values are considered: the policy π that relates the searcher's cell c with the action $\mathbb{a}$ to be chosen, and the value function V that defines the cumulated reward of the searcher.

In the probabilistic search described in the terms of Markov decision processes (MDP), policy defines the probability $\pi(\mathbb{a} \mid c)$ choosing the action $\mathbb{a}$ given that the searcher is in the cell c, and value function defines the estimated expected reward $V(c)$, which will be obtained by the searcher starting from the cell c.

In the actor-critic algorithms, definition of the policy and estimation of the expected reward are conducted by different agents: the actor deals with the policy $\pi(\mathbb{a}|c)$ aiming to maximize the expected reward $V(c)$, while estimation of the reward is conducted by the critic.

In the search process, the roles of the actor and the critic can be played either by the searcher, which in the case of search games they can be separated. Then the searcher acts as an actor, which maximizes the reward; while the target acts as a critic, which minimizes the searcher's reward.

Formally, the main idea of this approach is represented as follows. Assume that the policy π_θ of the searcher depends on a parameter θ and the critic estimated reward function V_ω depends on the parameter ω.

Denote by $J(\theta, \omega) \in \mathbb{R}$ an expected reward of the searcher obtained by using the policy π_θ with the parameter θ and the reward function V_ω with the parameter ω.

In addition, denote by $A(\mathbb{a}, c) \in \mathbb{R}$ an advantage of choosing action $\mathbb{a}$ in the cell c and by $Q_\omega(\mathbb{a}, c) \in \mathbb{R}$ a critic's estimation of the reward of choosing the action $\mathbb{a}$ in the cell c.

Then, the actor deals with the gradient of the function J with respect to the parameter θ

$$\nabla_\theta J(\theta, \omega) = \frac{1}{T}\sum_{t=1}^{T} \nabla_\theta \ln \pi_\theta(\mathbb{a}(t)|c(t))A(\mathbb{a}(t), c(t)),$$

and the critic deals with the gradient of the function J with respect to the parameter ω

$$\nabla_\omega J(\theta, \omega) = \frac{1}{T}\sum_{t=1}^{T} \nabla_\omega [Q_\omega(\mathbb{a}(t), c(t)) - V_\omega(c(t))]^2.$$

The actor's update of the parameter θ is defined by the gradient ascent

$$\theta' = \theta + \alpha \nabla_\theta J(\theta, \omega),$$

where α is a learning rate of the actor, and the critic's update of the parameter ω is defined by the gradient descent

$$\omega' = \omega - \beta \nabla_\omega J(\theta, \omega),$$

where β is a learning rate of the critic.

An advantage $A(\mathbb{a}(t), c(t))$ of choosing the action in the cell is usually defined by the difference

$$A(\mathbb{a}(t), c(t)) = Q_\theta(\mathbb{a}(t), c(t)) - V_\theta(c(t)),$$

where, as above, $Q_\theta(\mathbb{a}(t), c(t))$ is a cumulated reward of choosing action $\mathbb{a}(t)$ while being in the cell $c(t)$ and $V_\theta(c(t))$ is an expected reward obtained by the searcher starting from the cell $c(t)$.

Usually, learning of the parameters θ and ω is conducted using independent actor's and critic's neural networks; an example of the Python code can be found at the webpage [19].

Note that such an approach follows the line of general Turing systems [36] in which searcher acts as activator and target acts as inhibitor. Further considerations of these approaches are not in the scope of this book.

5.4 Conclusions

Usually, learning methods in the search and detection tasks are used for predicting the targets' locations based on the results of previous detections.

In the search for static targets, this goal is reached by updating the probability maps such that the already screened regions obtain zero targets' location probabilities and the probabilities of the targets' locations in the other regions are increased. The main problem in this case is to create an initial probability map, which is a crucial factor in the search for static targets.

In the search for moving targets, learning addresses the targets' mobility patterns that allow prediction of future targets' locations and planning the search, respectively. The main task in this search is to analyze the changes of the targets' location probabilities and, following these changes, to create a model of the targets' movements and then to predict the targets' locations.

In this chapter, we recalled main methods of machine learning used in the search problems. We started with the reinforcement learning that forms a basis for the algorithms considered in the next two chapters.

The use of supervised and unsupervised learning in the search and detection processes is illustrated by basic algorithms that stress the difference between these methods. Note that in the search and detection, supervised learning is effective only if there is enough prior information about the environmental conditions and the targets' behavior; otherwise, such learning has no serious advantages.

Finally, we considered two-actor search: the search games in which the target aims to escape from being detected and the actor-critic approach that extends the basic reinforcement learning method to two agents. The last approach inspired the use of separate neural networks for prediction and training, which is implemented in the algorithms considered in the next chapters.

6

Search with Deep Q-Learning: Single Agent

In this chapter, we continue the direction of heuristic algorithms for search and detection with false-positive and false-negative detection errors and consider the detection of static and moving targets. In addition, we assume that the agent is equipped with an on-board controller that is powerful enough to process deep Q-learning and train neural networks on relatively large data sets.

6.1 Problem Formulation and Notation

The implemented deep Q-learning scheme follows general deep learning techniques applied to search and detection processes and mobile agents navigation. In addition, this method utilizes knowledge about the targets' locations in the form of a probability map.

In the algorithm, it is assumed that the search agent starts with an initial probability map of the targets' locations and makes decisions about its further movements either by maximizing the expected cumulative information gain regarding the targets' locations or by minimizing the expected length of the agent's trajectory up to obtaining the desired probability map. For brevity, we refer to the first approach as the Q-max algorithm and the second approach as the shortest path length (SPL) algorithm.

Maximization of the expected information gain and minimization of the expected path length are performed with a conventional dynamic programming approach. At the same time, deep Q-learning of the appropriate neural network obtains the decision regarding the next step of the agent. As an input, the network receives the agent's location and current probability map and outputs the preferred move of the agent. Finally, a priori training of the network is conducted based on a set of simulated realizations of the considered detection process.

Let us recall the used terms and notation. By $C = \{c_1, c_2, \dots, c_n\}$ we denote a finite set of cells that represents a gridded two-dimensional domain, and it is assumed that in the domain C there are ξ targets, $\nu = 1, \dots, \xi$, $\xi \leq n - 1$, which can stay in their locations or move over the set C, and a search agent, which moves over the domain and is able to detect the targets by the appropriate sensors.

Following the Koopman approach, it is assumed that probability of detecting the target becomes higher as the agent moves closer to the target and as the agent observes the target's location for a longer time.

The goal is to find a policy for the agent's motion such that the agent detects all ξ targets in minimal time or, equivalently, it follows the shortest trajectory.

As above (see Section 3.1), the state $s(c_i, t)$ of each cell $c_i \in C$, $i = 1, 2, \ldots, n$, at time $t = 1, 2, \ldots$ is considered a random variable with the values $s(c_i, t) \in \{0, 1\}$, where $s(c_i, t) = 0$ means that the cell c_i at time t is empty, and $s(c_i, t) = 1$ means that the cell c_i at time t is occupied by a target. The probabilities that the cell is occupied and is empty are complementary, so

$$Pr\{s(c_i, t) = 0\} + Pr\{s(c_i, t) = 1\} = 1.$$

The probability of detecting the target is defined by the Koopman exponential random search formula [39]

$$\Pr\{\text{target detected in } c_i \mid \text{target located in } c_i\} = 1 - \exp[-\kappa(dist(c_i, c_j), \tau)],$$

where, as above, $\kappa(dist(c_i, c_j), \tau)$ is the search effort applied to the cell c_i given the distance $dist(c_i, c_j)$ and observation period τ.

The possibility of false-positive and false-negative detection errors is represented by true and false alarms that can be broadcasted by the occupied cells:

- the cells with the states $s(c, t) = 1$ broadcast alarm $\tilde{a}(c, t) = 1$ with probability

$$p_{\text{TA}} = \Pr\{\tilde{a}(c, t) = 1 \mid s(c, t) = 1\},$$

- the cells with the states $s(c, t) = 0$ broadcast the alarm $\tilde{a}(c, t) = 1$ with probability

$$p_{\text{FA}} = Pr\{\tilde{a}(c, t) = 1 \mid s(c, t) = 0\} = \alpha p_{TA},$$

where $0 \leq \alpha < 1$.

We assume that the probability of perceiving alarm is also defined by the Koopman formula

$$\Pr\{\text{alarm percieved at } c_j \mid \text{alarm sent from } c_i\} = \exp[-d(c_i, c_j)/\lambda],$$

where λ is the sensitivity of the sensor installed on the agent; it is assumed that all the cells are observed during the same period, so the value τ can be omitted.

Let $p_i(t) = \Pr\{s(c_i, t) = 1\}$ be a probability that at time t cell c_i, $i = 1, 2, \ldots, n$, is occupied by a target and let

$$P(t) = \{p_1(t), p_2(t), \ldots, p_n(t)\}$$

be a probability map of the set C.

Probability of the event $\tilde{x}_j(c_i, t)$, $i, j = 1, 2, \ldots, n$, that at time t the search agent located in cell c_j receives an alarm from cell c_i, is

$$p(\tilde{x}_j(c_i, t)) = p_i(t-1)p_{TA}\exp[-d(c_i, c_j)/\lambda] + (1 - p_i(t-1))p_{FA}\exp[-d(c_i, c_j)/\lambda],$$

and the probability of the event $\overline{\tilde{x}_j(c_i, t)}$, that this agent does not receive an alarm at time t, is

$$p(\overline{\tilde{x}_j(c_i, t)}) = 1 - p(\tilde{x}_j(c_i, t)).$$

Note that the event $\tilde{x}_j(c_i, t)$ represents the fact that the agent does not distinguish between true and false alarms, but it indicates that the agent receives a signal (which can be either a true or false alarm) from cell c_i. If $\alpha = 1$ and therefore $p_{\text{TA}} = p_{\text{FA}}$, then

$$p(\tilde{x}_j(c_i, t)) = p_{\text{TA}}\exp[-d(c_i, c_j)/\lambda],$$

which means that the agent's knowledge about the targets' locations does not depend on the targets' location probabilities $p_i(t)$, $i = 1, 2, \ldots, n$.

When the search agent located in cell c_j receives an alarm from cell c_i, the probability that cell c_i is occupied by the target is

$$\Pr\{s(c_i,t)=1 \mid \tilde{x}_j(c_i,t)\} = \frac{p_i(t-1)Pr\{\tilde{x}_j(c_i,t) \mid s(c_i,t)=1\}}{p_i(t-1)Pr\{\tilde{x}_j(c_i,t) \mid s(c_i,t)=1\} + (1-p_i(t-1))Pr\{\tilde{x}_j(c_i,t) \mid s(c_i,t)=0\}},$$

and the probability that c_i is occupied by the target when the agent does not receive an alarm from c_i is

$$\Pr\{s(c_i,t)=1 \mid \overline{\tilde{x}_j(c_i,t)}\} = \frac{p_i(t-1)Pr\{\overline{\tilde{x}_j(c_i,t)} \mid s(c_i,t)=1\}}{p_i(t-1)Pr\{\overline{\tilde{x}_j(c_i,t)} \mid s(c_i,t)=1\} + (1-p_i(t-1))Pr\{\overline{\tilde{x}_j(c_i,t)} \mid s(c_i,t)=0\}},$$

where probabilities $p_i(t-1)$, $i = 1, 2, \ldots, n$, represent the agent's knowledge about the targets' locations at time $t-1$, and it is assumed that the initial probabilities $p_i(0)$ at time $t = 0$ are defined with respect to prior information; if there is any initial information about the targets' locations, it is assumed that $p_i(0) = \frac{1}{2}$ for each $i = 1, 2, \ldots, n$.

In the framework of the Koopman approach, these probabilities are

$$\Pr\{s(c_i,t)=1 \mid \tilde{x}_j(c_i,t)\} = \frac{p_i(t-1)p_{\mathrm{TA}}}{p_i(t-1)p_{\mathrm{TA}} + (1-p_i(t-1))\alpha p_{\mathrm{TA}}},$$

and

$$\Pr\{s(c_i,t)=1 \mid \overline{\tilde{x}_j(c_i,t)}\} = \frac{p_i(t-1)(1-p_{\mathrm{TA}}\exp[-d(c_i,c_j)/\lambda])}{p_i(t-1)(1-p_{\mathrm{TA}}\exp[-d(c_i,c_j)/\lambda]) + (1-p_i(t-1))(1-\alpha p_{\mathrm{TA}}\exp[-d(c_i,c_j)/\lambda])}.$$

The defined process of updating the probabilities with respect to the received alarms is illustrated in Figure 6.1.

As above, target detection is processed as follows. At each step, the agent considers the probabilities $p_i(t)$, $i = 1, 2, \ldots, n$, of the targets' locations and makes a decision regarding its next movement. After moving to the new location (or remaining at its current location), the agent receives alarms from the available cells and updates the probabilities $p_i(t)$. The obtained updated probabilities $p_i(t+1)$ are used to continue the process.

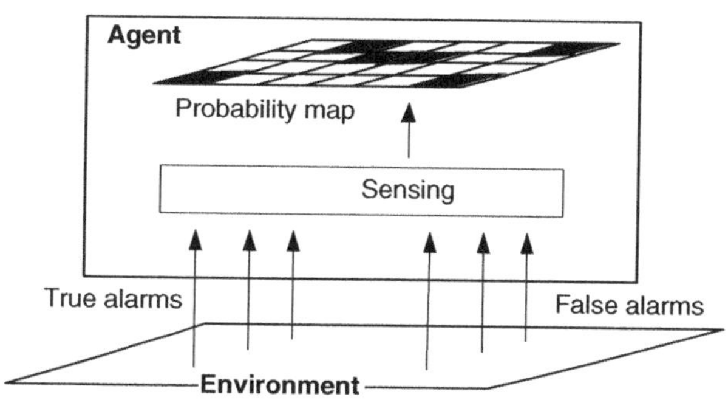

Figure 6.1 Receiving information and updating the probability map.

6.2 Decision-making Policy and Deep Q-Learning Solution

The policy of choosing the action $\mathbb{a}(t)\colon C \to C$ which specifies to which cell the search agent moves depends on the searchers' knowledge about targets' locations. The goal is to define a policy such that the agent detects all ξ targets in minimal time.

6.2.1 Agent's Actions and Decision-making

We assume that the policy $\pi\colon C \times P \to \mathbb{a}$ for choosing an action does not depend on time and is specified for any t by the current agent's location $c(t)$ and probability map $P(t)$. Then, the desired policy should produce actions such that the agent's trajectory from the cell $c(0)$ up to the final cell $c(T)$ is as short as possible, and by following this trajectory, the agent detects all ξ targets. It is assumed that the number ξ of targets is unavailable to the agent during detection and is used to indicate the end of the detection process.

With respect to the indicated properties of the desired agent trajectory, a search for the decision-making policy can follow either the maximization of the expected cumulative information gain over the trajectory or the direct optimization of the trajectory length in the indicated sense of minimal detection time. The first approach is referred to as the Q-max algorithm, and the second is referred to as the SPL algorithm.

In the previous chapters, we considered a similar search and detection problem that was solved heuristically by evaluating the decisions made at each step of the search process. Here we address a more sophisticated approach that implements deep Q-learning techniques. We consider the information-based Q-max algorithm and then the SPL algorithm, which deals directly with the shortest path length (SPL).

Let us start with the Q-max solution. Assume that at each time t the agent is located in cell $c(t)$ and the action $\mathbb{a}(t)$ is chosen from the possible movements

$$\mathbb{a}(t) \in \mathbb{A} = \{\uparrow, \nearrow, \rightarrow, \searrow, \downarrow, \swarrow, \leftarrow, \nwarrow, \odot\},$$

where, as in Section 4.1, arrows define the directions of the steps and $\odot$ means "to stay in the current position".

Denote by $P_{\mathbb{a}}(t+1)$ a probability map that should represent the targets' locations at time $t+1$ given that at time t, the agent chooses action $\mathbb{a}(t)$. Then, given action $\mathbb{a}_t$, the immediate expected informational reward of the agent is defined by the Kullback–Leibler distance

$$R(\mathbb{a}, t) = D_{\mathrm{KL}}(P_{\mathbb{a}}(t+1) || P(t))$$

between the map $P_{\mathbb{a}}(t+1)$ and the current probability map $P(t)$.

Given a policy π for choosing an action, the expected cumulative discounted reward obtained by an agent that starts in cell $c(t)$ with probability map $P(t)$ and chooses action $\mathbb{a}(t)$ is

$$q_{\pi}(c(t), P(t), \mathbb{a}(t)) = \mathbb{E}_{\pi}\left\{\sum_{\tau=0}^{\infty} \gamma^{\tau} R(\mathbb{a}, t+\tau)\right\},$$

where $\gamma \in (0, 1]$ is the discount factor. The goal is to find a maximum value

$$Q(c(t), P(t), \mathbb{a}(t)) = \max_{\pi} q_{\pi}(c(t), P(t), \mathbb{a}(t))$$

of the expected reward q_{π} over all possible policies π that can be applied after action $\mathbb{a}(t)$ at time t.

Since the number of possible policies is infinite, the value $Q(c(t), P(t), \mathbb{a}(t))$ of the maximal expected cumulative discounted reward cannot be calculated exactly, and for any realistic scenario, it should be approximated. Below, we follow the deep Q-learning approach and consider the Q-max algorithm, which approximates the values $Q(c(t), P(t), \mathbb{a}(t))$ of the reward for all possible actions and therefore provides criteria for choosing the actions.

6.2.2 Dynamic Programming and Target Neural Networks

In the considered Q-max algorithm, the learning stage is based on a neural network for predicting current rewards. In the simplest configuration, which is still rather effective, the network consists of one input layer, which includes $2n$ neurons (recall that n is the size of the domain); one hidden layer, which also includes $2n$ neurons; and an output layer, which includes $\#\mathbb{A} = 9$ neurons with respect to the number of possible actions.

The inputs of the network are defined as follows. The first chunk of n inputs $(1, 2, \ldots n)$ receives a binary vector that represents the agent's location. Namely, if the agent is located in cell c_j, then the jth input of the network is equal to 1 and the other $n-1$ inputs are equal to 0. The second chunk of n inputs $(n+1, n+2, \ldots 2n)$ receives the targets' location probabilities; namely, the $(n+i)$th input receives the target's location probability p_i, $i = 1, 2, \ldots, n$, as it appears in the probability map P.

Hidden layer of the network consists of $2n$ neurons, each of which implements a fully connected linear layer and sigmoid activation function

$$f(x) = 1/(1 + e^{-x}).$$

Such activation function was chosen among several possible activation functions (e.g. step function, Soft plus function and SiLU function) and was found to provide effective learning in all conducted simulations.

Nine neurons of the output layer correspond to the possible actions. Namely, the first output corresponds to the action $\mathbb{a}_1 =$ " $\uparrow$ ", "step forward"; the second output corresponds to the action $\mathbb{a}_2 =$ " $\nearrow$ ", "step right-forward"; and so on up to action $\mathbb{a}_9 =$ " $\odot$ ", which is "stay in the current cell." The value of the kth output is the maximal expected cumulative discounted reward $Q(c_j, P, \mathbb{a}_k)$ obtained by the agent if it is in cell c_j, $j = 1, 2, \ldots, n$, and given the target location probabilities $P = (p_1, p_2, \ldots, p_n)$, it chooses action $\mathbb{a}_k$, $k = 1, 2, \ldots, 9$.

The scheme of the network is shown in Figure 6.2.

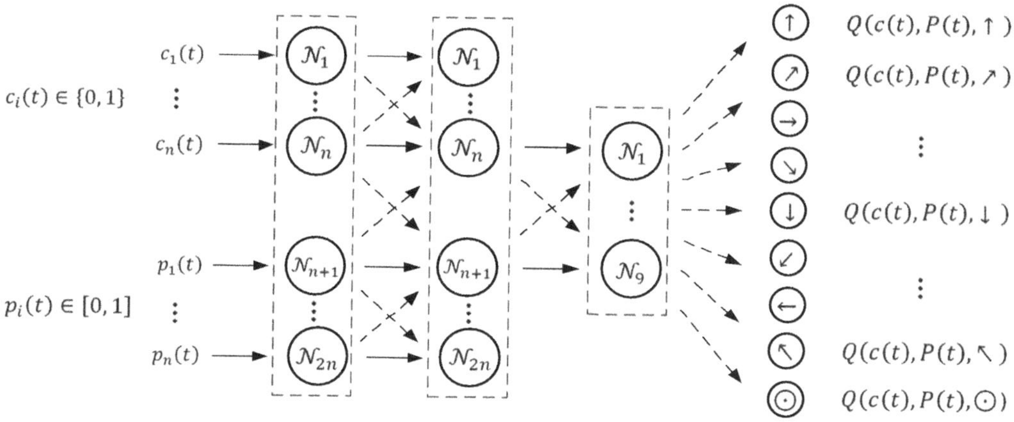

Figure 6.2 The scheme of neural network used in the learning stage of the Q-max algorithm.

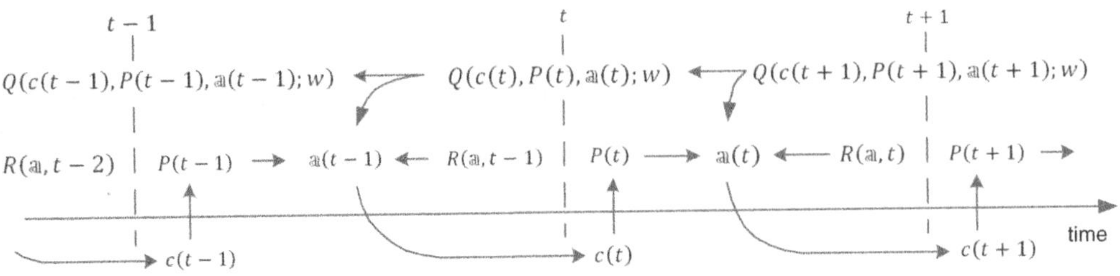

Figure 6.3 The scheme of data flow in the training stage of the network.

Training stage of the network implements deep Q-learning techniques, which follow the dynamic programming approach. The Bellman equation for calculating maximal cumulative discounted reward is as follows:

$$Q(c(t), P(t), \mathbb{a}(t)) = R(\mathbb{a}, t) + \gamma \max_{\mathbb{a} \in \mathbb{A}} Q(c(t+1), P(t+1), \mathbb{a}), \qquad (*)$$

and this equation forms a basis for updating the weights of the links in the network. Data flow specified by this equation is shown in Figure 6.3.

Let w be a vector of the link weights of the network. In the considered case, there are $4n^2 + 18n + 2n + 9$ values of the weights, where $4n^2$ is the number of links between the input layers and the hidden layer, $18n$ is the number of links between the hidden layer and the output layer, $2n$ is the number of biases in the hidden layer and 9 is the number of biases in the output layer.

In addition, to distinguish these steps and to separate them from the real time, below we enumerate the training steps by $l = 1, 2, \ldots$ and retain the symbol t for the real time of the detection process.

Denote by $Q(c(l), P(l), \mathbb{a}(l); w)$ maximal cumulative discounted reward calculated at step $l = 1, 2, \ldots$ by the network with weights w, and denote by $Q^+(c(l), P(l), \mathbb{a}(l); w')$ the expected maximal cumulative discounted reward calculated using the vector w' of the updated weights following the recurrent equation (*). That is

$$Q^+(c(l), P(l), \mathbb{a}(l); w') = R(\mathbb{a}, l) + \gamma \max_{\mathbb{a} \in \mathbb{A}} Q(c(l+1), P(l+1), \mathbb{a}; w')$$

Then, the temporal difference learning error is

$$\Delta_l(Q) = Q^+(c(l), P(l), \mathbb{a}(l); w') - Q(c(l), P(l), \mathbb{a}(l); w)$$

In practice, the values $Q(c(l), P(l), \mathbb{a}(l); w)$ and $Q^+(c(l), P(l), \mathbb{a}(l); w')$ are associated with separate neural networks; the first is called the prediction network, and the second is called the target network.

6.2.3 Model-Free and Model-Based Learning

The updating of the weights w in the prediction network is conducted by the following basic back-propagation techniques. The weights for the next step $l+1$ are updated with respect to the temporal difference learning error $\Delta_l(Q)$ calculated at the current step l. The weights w' in the target network are updated to the values of the weights w after an arbitrary number of iterations. In the simulations presented in next section, such updating was conducted at every fifth step.

The training procedure uses the events occurring in the environment and does not require prior knowledge about the targets' abilities. In other words, following this procedure the agent detects the targets in the environment, simultaneously learns the environment and trains the

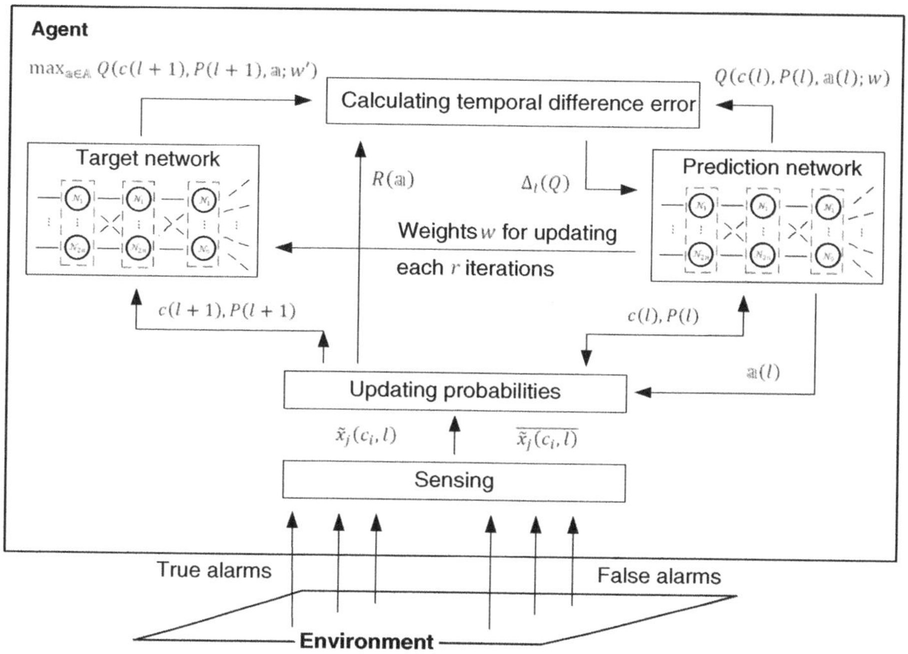

Figure 6.4 Actions of the online model-free learning in the Q-max algorithm.

neural network that supports the agent's decision-making processes. We refer to such a scenario as *model-free learning*.

Actions of online model-free learning are illustrated in Figure 6.4.

Following the figure, at step l, the target location probabilities $\Pr\{s(c_i, l) = 1 \mid \tilde{x}_j(c_i, l)\}$ and $\Pr\{s(c_i, l) = 1 \mid \overline{\tilde{x}_j(c_i, l)}\}$, $i, j = 1, 2, \ldots, n$, are updated with respect to the events $\tilde{x}_j(c_i, l)$ and $\overline{\tilde{x}_j(c_i, l)}$ of receiving and not receiving a signal from cell c_i while the agent is in cell c_j.

The updated target location probabilities are used for calculating the value of the immediate reward $R(a, l)$ and the value $Q^+(c(l), P(l), a(l); w')$ in the target network.

In parallel, the value $Q(c(l), P(l), a(l); w)$ of the prediction network is used for choosing the action and, consequently, specifying the agent's expected position in the environment.

After calculating the temporal difference error $\Delta_l(Q)$ between the Q-values in the target and in the prediction network, the weights w in the prediction network are updated, and the process continues with step $l+1$.

Note that in all given above definitions, cumulative reward does not depend on the previous trajectory of the agent. Hence, the process that governs the agent's activity is a Markov process with states that include positions of the agent and corresponding probability maps. This property allows the use of an additional offline learning procedure based on the knowledge of the targets' abilities.

If the abilities of the targets are known and can be represented in the form of transition probability matrices that govern the targets' motion, the learning process can be conducted offline without checking the events occurring in the environment. In this scenario, at step l, instead of the target location probabilities $\Pr\{s(c_i, l) = 1 \mid \tilde{x}_j(c_i, l)\}$ and $\Pr\{s(c_i, l) = 1 \mid \overline{\tilde{x}_j(c_i, l)}\}$, $i, j = 1, 2, \ldots, n$, the networks use the probabilities of the expected targets' locations $\Pr\{s(c_i, l) = 1 \mid s(c_i, l-1) = 1\}$ and $\Pr\{s(c_i, l) = 1 \mid s(c_i, l-1) = 0\}$ at step l given the states of the cells at the previous step $l-1$.

Based on the previous definitions, these probabilities are defined as follows:

$$\Pr\{s(c_i, l) = 1 \mid s(c_i, l-1) = 1\}$$
$$= \frac{p_i(l-1)p_{\mathrm{TA}}\exp[-d(c_i, c_j)/\lambda]}{p_i(l-1)(1-\alpha)+\alpha}$$
$$+ \frac{p_i(l-1)(1-\exp[-d(c_i, c_j)/\lambda])^2}{p_i(l-1)(1-p_{\mathrm{TA}}\exp[-d(c_i, c_j)/\lambda]) + (1-p_i(l-1))(1-\alpha p_{\mathrm{TA}}\exp[-d(c_i, c_j)/\lambda])}$$

and

$$\Pr\{s(c_i, l) = 1 \mid s(c_i, l-1) = 0\}$$
$$= \frac{p_i(l-1)\alpha p_{\mathrm{TA}}\exp[-d(c_i, c_j)/\lambda]}{p_i(l-1)(1-\alpha)+\alpha}$$
$$+ \frac{p_i(l-1)(1-\exp[-d(c_i, c_j)/\lambda])(1-\alpha p_{\mathrm{TA}}\exp[-d(c_i, c_j)/\lambda])}{p_i(l-1)(1-p_{\mathrm{TA}}\exp[-d(c_i, c_j)/\lambda]) + (1-p_i(l-1))(1-\alpha p_{\mathrm{TA}}\exp[-d(c_i, c_j)/\lambda])}.$$

Since the presented procedure is based on certain knowledge about the targets' activity, it is called the *model-based learning* procedure.

The actions of the model-based offline learning procedure are illustrated in Figure 6.5.

The model-based learning procedure differs from the model-free learning procedure only in using the target location probabilities and the method of updating them. In the model-free procedure, these probabilities are specified based on the events occurring in the environment. In the model-free procedure, they are calculated by following the Markov property of the system without referring to the real events in the environment.

As a result, in the model-free procedure learning is slower than in the model-based procedure. However, while in the first case, the agent learns during the detection process and can act without prior information about the targets' abilities, in the second case, it starts detection only after offline learning and requires an exact model of the targets' activity. Thus, the choice of a particular procedure is based on the considered practical task and available information.

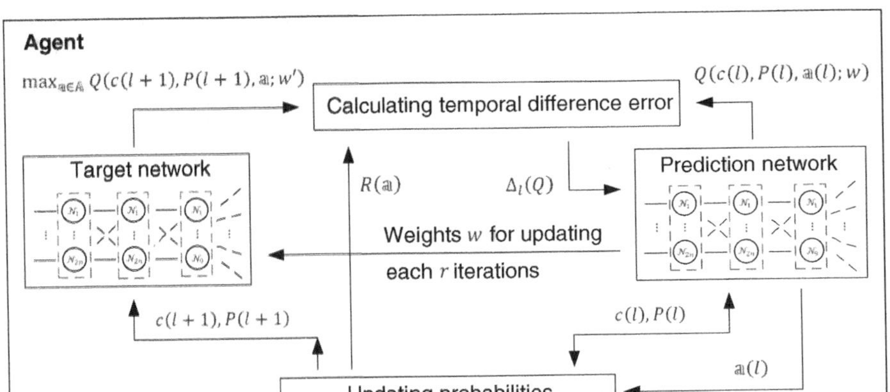

Figure 6.5 Actions of the offline model-based learning in the Q-max algorithm.

6.2.4 Choice of the Actions at the Learning Stage

As indicated above, given the agent's location $c(l)$ and the targets' probability map $P(l)$, the neural networks used at the learning stage provide *nine* output Q-values that are associated with possible actions $\mathbb{a}_k \in \mathbb{A}$, $k = 1, 2, \dots, 9$, namely,

$$Q(c(l), P(l), \mathbb{a}_1; w), Q(c(l), P(l), \mathbb{a}_2; w), \dots, Q(c(l), P(l), \mathbb{a}_9; w)$$

where $\mathbb{a}_1 =$ " $\uparrow$ " ("step forward"), $\mathbb{a}_2 =$ " $\nearrow$ " ("step right-forward"), and so on up to $\mathbb{a}_9 =$ " $\odot$ " ("stay in the current cell").

The choice among the actions $\mathbb{a}_k \in \mathbb{A}$ is based on the corresponding Q-values $Q(c(l), P(l), \mathbb{a}_k; w)$, $k = 1, 2, \dots, 9$, and implements exploration and exploitation techniques. At the initial step $l = 0$, when the agent has no prior learned information about the targets' locations, action $\mathbb{a}(l) \in \mathbb{A}$ is chosen randomly.

Then, after processing the step prescribed by action $\mathbb{a}(l)$, the next action $\mathbb{a}(l+1)$ is chosen either based on the target location probabilities learned by the neural networks or randomly from among the actions available at this step.

The ratio of random choices decreases with the number of steps, and after finalizing the learning processes in the neural networks, the actions are chosen with respect to the Q-values only.

Formally, this process can be defined using different policies, for example, with a decaying ε-greedy policy that uses the probability ε, which decreases with the increase in the number of steps from its maximal value $\varepsilon = 1$ to the minimal value $\varepsilon = 0$.

The agent chooses an action randomly with probability ε and according to the greedy rule $\arg\max_{\mathbb{a} \in \mathbb{A}} Q(c(l), P(l), \mathbb{a}; w)$ with probability $1 - \varepsilon$. In this policy, the choice of action is governed by the probability ε and does not depend on the Q-values of the actions.

A more sophisticated policy of intermittence between random and greedy choice is the SoftMax policy. In this policy, the probability $p(\mathbb{a}_k \mid Q; \eta)$ of choosing action $\mathbb{a}_k$ is defined with respect to both the parameter $\eta \in [0, +\infty)$ and the Q-values of the actions

$$p(\mathbb{a}_k \mid Q; \eta) = \frac{\exp[Q(c(l), P(l), \mathbb{a}_k; w)/\eta]}{\sum_{j=1}^{9} \exp[Q(c(l), P(l), \mathbb{a}_j; w)/\eta]}.$$

Therefore, if $\eta \to 0$, then $p(\mathbb{a}_k \mid Q; \eta) \to 1$ for $\mathbb{a}_k = \arg\max_{\mathbb{a} \in \mathbb{A}} Q(c(l), P(l), \mathbb{a}; w)$ and $p(\mathbb{a}_k \mid Q; \eta) \to 0$ for all other actions, and if $\eta \to \infty$, then $p(\mathbb{a}_k \mid Q; \eta) \to \frac{1}{9}$, which corresponds to a randomly chosen action.

The intermediate values $0 < \eta < \infty$ correspond to the probabilities $p(\mathbb{a} \mid Q; \eta) \in (0, 1)$ and govern the randomness of the action choice. In other words, the value of the parameter η decreases with the increasing number of steps l from its maximal value to zero; thus, for the unlearned networks, the agent chooses actions randomly and then follows the information about the targets' locations learned by the networks.

The first stages with randomly chosen actions are usually interpreted as exploration stages, and the later stages, based on the learned information, are considered exploitation stages. In the simulations, we considered both policies and finally implemented the SoftMax policy since it provides more correct choices, especially in cases with relatively high Q-values associated with different actions.

6.2.5 The Q-max Algorithm

The Q-max algorithm can be used both with model-free learning for direct online detection and with model-based learning for offline policy planning and further online applications of this policy.

Since both learning processes follow the same steps (with the only difference in the source of information about the targets' location probabilities), below, we outline the Q-max algorithm with model-based learning.

The Q-max algorithm with model-based learning includes three stages: at the first stage, the algorithm generates the training data set, which includes reasonable probability maps with possible agent locations; at the second stage, it trains the prediction neural network using the generated data set; and at the third stage, the algorithm solves detection problem by following the decisions made by the trained prediction neural network.

The first stage of generating the training data set is outlined as follows.

Algorithm 6.1 Generating the training data set for Q-max network

---

Input: Set $C = \{c_1, c_2, \dots, c_n\}$,
set $\mathbb{A} = \{\uparrow, \nearrow, \rightarrow, \searrow, \downarrow, \swarrow, \leftarrow, \nwarrow, \odot\}$ of possible movements,
probability p_{TA} of true alarms,
rate α of false alarms and probability $p_{\mathrm{FA}} = \alpha p_{\mathrm{TA}}$,
sensor sensitivity λ,
range $[\xi_1, \xi_2]$ of possible numbers $0 < \xi_1 < \xi_2 \leq n - 1$ of the targets,
length $L \in (0, \infty)$ of the agent's trajectory,
number $N \in (0, \infty)$ of agent trajectories,
initial probability map $P(0) = (p_1(0), p_2(0), \dots, p_n(0))$ on the set C.

Output: Data set: an $L \times N$ table of pairs (c, P) of agent positions c and probability maps P

1. Create $L \times N$ data table.
2. For each agent trajectory $j = 1, \dots, N$ do:
3. Choose a number $\xi \in [\xi_1, \xi_2]$ of targets according to a uniform distribution.
4. Choose target locations $c_1, c_2, \dots, c_\xi \in C$ according to the uniform distribution.
5. Choose initial agent position $c(0) \in C$ randomly according to the uniform distribution.
6. For $l = 0, \dots, L - 1$ do:
7. Save the pair $\langle c(l), P(l) \rangle$ as the jth element of the data table.
8. Choose an action $\mathbb{a}(l) \in \mathbb{A}$ randomly according to the uniform distribution.
9. Set $c(l + 1) = \mathbb{a}(c(l))$.
Set probabilities $p_i(l + 1)$, $i = 1, 2, \dots, n$, for the probability map $P(l + 1)$:
10. If $s(c_i, l) = 1$, then

$$p_i(l+1) = \frac{p_i(l) p_{\mathrm{TA}} \exp[-d(c_i, c_j)/\lambda]}{p_i(l)(1-\alpha)+\alpha} + \frac{p_i(l)(1-\exp[-d(c_i, c_j)/\lambda])^2}{p_i(l)(1 - p_{\mathrm{TA}} \exp[-d(c_i, c_j)/\lambda]) + (1-p_i(l))(1 - \alpha p_{\mathrm{TA}} \exp[-d(c_i, c_j)/\lambda])},$$

11. Else

$$p_i(l+1) = \frac{p_i(l) \alpha p_{\mathrm{TA}} \exp[-d(c_i, c_j)/\lambda]}{p_i(l)(1-\alpha)+\alpha} + \frac{p_i(l)(1-\exp[-d(c_i, c_j)/\lambda])(1 - \alpha p_{\mathrm{TA}} \exp[-d(c_i, c_j)/\lambda])}{p_i(l)(1 - p_{\mathrm{TA}} \exp[-d(c_i, c_j)/\lambda]) + (1-p_i(l))(1 - \alpha p_{\mathrm{TA}} \exp[-d(c_i, c_j)/\lambda])},$$

12. End if.
13. End for.
14. End for.
15. Return data table.

Training data set includes N random trajectories of length L; each element of the data set is a pair of the position of search agent and corresponding probability map.

Note that in the generated data set, the agent's locations are taken from the connected trajectory and the corresponding probability maps are calculated with respect to these positions, possible actions, sensing abilities, and environmental conditions. Hence, this generated data can be considered a good imitation of real data and can be used at the learning stage of the prediction network

The generated agent locations and corresponding probability maps are used as input of the prediction neural network in the training stage. The goal of the training is specified by the objective probability map $P^* = \{p_1^*, p_2^*, \ldots, p_n^*\}$, which defines the target location probabilities that provide sufficient information for immediately detecting all the targets.

In the best case, we have probabilities $p_i^* \in \{0, 1\}$, and in practical scenarios, it is assumed that either $p_i^* \in [0, \varepsilon]$ or $p_i^* \in [1 - \varepsilon, 1]$ for certain $0 < \varepsilon \ll 1$, $i = 1, 2, \ldots, n$.

The training stage of the Q-max algorithm is outlined as follows.

Algorithm 6.2 Training the prediction network

Network structure:

input layer: $2n$ neurons (n agent positions and n target location probabilities with respect to the size n of the set C),
hidden layer: $2n$ neurons,
output layer: 9 neurons (with respect to the number of actions – size of the set $\mathbb{A}$).

Activation function:

sigmoid function $f(x) = 1/(1 + e^{-x})$.

Loss function:

mean square error (MSE) function.

Input: Set $C = \{c_1, c_2, \ldots, c_n\}$,
set $\mathbb{A} = \{\uparrow, \nearrow, \rightarrow, \searrow, \downarrow, \swarrow, \leftarrow, \nwarrow, \odot\}$ of possible movements,
probability p_{TA} of true alarms,
rate α of false alarms and probability $p_{FA} = \alpha p_{TA}$,
sensor sensitivity λ,
discount factor γ,
value ε,
objective probability map P^* (obtained by using the value ε),
number r of iterations for updating the weights,
initial value η and discount factor δ,
learning rate ρ (with respect to the type of optimizer),
number M of epochs,
initial weights w of the prediction network and initial weights $w' = w$ of the target network,
training data set ($L \times N$ table of (c, P) pairs created by Algorithm 6.1).

Output: Data set: an $L \times N$ table of pairs (c, P) of agent positions c and probability maps P

1. Create the prediction network.
2. Create the target network as a copy of the prediction network.
3. For each epoch $j = 1, \ldots, M$ do:
4. For each pair (c, P) from the training data set, do:
5. For each action $\mathbb{a} \in \mathbb{A}$ do:

6. Calculate value $Q(c, P, \mathbb{a}; w)$ with the prediction network.
7. Calculate probability

$$p(\mathbb{a}_k|Q;\eta) = \frac{\exp[Q(c, P, \mathbb{a}_k; w)/\eta]}{\sum_{j=1}^{9} \exp[Q(c, P, \mathbb{a}_j; w)/\eta]}.$$

8. End for.
9. Choose an action $\mathbb{a}$ according to the probabilities $p(\mathbb{a}|Q;\eta)$.
10. Set $c' = \mathbb{a}(c)$.
    Set probabilities p_i' for the probability map P' using the probabilities p_i from the map P, $i = 1, 2, \dots, n$:
11. If $s(c_i, l) = 1$, then

$$p_i' = \frac{p_i p_{\mathrm{TA}} \exp[-d(c_i, c_j)/\lambda]}{p_i(1-\alpha)+\alpha} + \frac{p_i(1-\exp[-d(c_i, c_j)/\lambda])^2}{p_i(1-p_{\mathrm{TA}}\exp[-d(c_i, c_j)/\lambda]) + (1-p_i)(1-\alpha p_{\mathrm{TA}}\exp[-d(c_i, c_j)/\lambda])},$$

12. Else

$$p_i' = \frac{p_i \alpha p_{\mathrm{TA}} \exp[-d(c_i, c_j)/\lambda]}{p_i(1-\alpha)+\alpha} + \frac{p_i(1-\exp[-d(c_i, c_j)/\lambda])(1-\alpha p_{\mathrm{TA}}\exp[-d(c_i, c_j)/\lambda])}{p_i(1-p_{\mathrm{TA}}\exp[-d(c_i, c_j)/\lambda]) + (1-p_i)(1-\alpha p_{\mathrm{TA}}\exp[-d(c_i, c_j)/\lambda])},$$

13. End if.
14. If $P = P^*$ or $c' \notin C$, then
15. Set immediate reward $R(\mathbb{a}) = 0$.
16. Else
17. Calculate immediate reward $R(\mathbb{a})$ with respect to P and P'

$$R(\mathbb{a}) = D_{KL}(P'\|P).$$

18. End if.
19. For each action $\mathbb{a} \in \mathbb{A}$ do:
20. If $P = P^*$ then
21. Set $Q(c', P', \mathbb{a}; w') = 0$,
22. Else
23. Calculate the value $Q(c', P', \mathbb{a}; w')$ with the target network,
24. End if.
25. End for.
26. Calculate target value

$$Q^+ = R(\mathbb{a}) + \gamma \max_{\mathbb{a}\in\mathbb{A}} Q(c', P', \mathbb{a}; w').$$

27. Calculate the temporal difference learning error for the chosen action $\mathbb{a}$

$$\Delta_l(Q) = Q^+ - Q(c, P, \mathbb{a}; w)$$

28. Set $\Delta_l(Q) = 0$ for all other actions.
29. Update the weights w in the prediction network by backpropagation with respect to the error $\Delta_l(Q)$.
30. Every r iterations, set the weights of the target network as $w' = w$.
31. End for.
32. End for.
33. Return: trained prediction network.

The network was validated using a data set that included the pairs (c, P) that are similar to the pairs appearing in the training data set but were not used in the training procedure. The size of the validation data set is approximately ten percent of the size of the training data set.

After training, the Q-max algorithm can be applied to simulated data or in a real search. The algorithm's structure mimics the search conducted by rescue and military services: first, the algorithm learns about the environment (by itself or at least using the model). Then it continues with the search in the real environment, where the probability map is updated with respect to the received alarms and acquired events, and decisions are made using the prediction network.

6.2.6 The SPL Algorithm

Now let us consider the SPL algorithm. Formally, it follows the same ideas and implements the same approach as the Q-max algorithm. The main difference is in the definition of the goal function. In the SPL algorithm, the goal function directly represents the agent's aim to detect all targets in minimal number of steps or, which is the same, to take minimal number of actions before reaching the termination condition.

In parallel to the reward $R(\mathbb{a}, t)$ of the action $\mathbb{a}(t) \in \mathbb{A}$ conducted at time t, in the SPL algorithm we define the penalty or the price paid by the search agent for action $\mathbb{a}(t) \in \mathbb{A}$. In consideration of the SPL, the payoff corresponds to the steps of the agent. We assume that all steps have equal unit price

$$O(\mathbb{a}, t) = 1$$

for each time $t = 1, 2, \ldots$ until termination of the search. Note that even if the agent chooses to stay at its current position, the payoff is calculated as 1.

Then, given a policy π for choosing an action, the expected cumulative payoff $pl_\pi(c(t), P(t), \mathbb{a}(t))$ of an agent that starts in cell $c(t)$ with probability map $P(t)$ and chooses action $\mathbb{a}(t)$ is

$$pl_\pi(c(t), P(t), \mathbb{a}(t)) = \mathbb{E}_\pi \left\{ \sum_{\tau=0}^{\infty} O(\mathbb{a}, t + \tau) \right\}$$

and the goal is to find the minimum value

$$\mathrm{SPL}(c(t), P(t), \mathbb{a}(t)) = \min_\pi pl_\pi(c(t), P(t), \mathbb{a}(t))$$

of the expected payoff pl_π over all possible policies π that can be applied after action $\mathbb{a}(t)$ is chosen at time t.

Then, the Bellman equation for calculating the defined minimal expected path length is

$$\mathrm{SPL}(c(t), P(t), \mathbb{a}(t)) = O(\mathbb{a}, t) + \min_{\mathbb{a} \in \mathbb{A}} \mathrm{SPL}(c(t + 1), P(t + 1), \mathbb{a})$$

and equations that define the training and functionality of the neural networks follow this equation and have the same form as in the Q-max algorithm (with the obvious substitution of maximization by minimization and the use of $\gamma = 1$).

6.3 Numerical Simulations

The presented algorithms were implemented and tested in several scenarios, and their functionality was compared with the functionality of previously developed heuristic algorithms and, in certain simple setups, with the algorithm that provides the optimal solution. Numerical simulations include training the neural network, simulation of the detection process by Q-max and SPL algorithms, and their comparisons with heuristic and optimal solutions.

Numerical simulations were implemented using basic tools of the Python programming language with the PyTorch machine learning library, and the trials were run on a PC Intel® Core™ i7-10700 CPU with 16 GB RAM.

In the simulations, detection was conducted over a gridded square domain of size $n = n_1 \times n_2$ cells, and it was assumed that the agent and each target could occupy only one cell. Given this equipment, we measured the run-time of the simulations for different datasets, which demonstrated that the algorithms are implementable on usual computers and do not require specific apparatus for their functionality.

6.3.1 Network Training in the Q-max Algorithm

First, let us consider simulation of the network training. The purpose of these simulations is to verify the training method and demonstrate a decrease in the temporal difference learning error $\Delta(Q)$ with an increasing number of learning epochs. Since the network training is the same for both the Q-max and SPL algorithms, we consider the training for the Q-max algorithm.

The training data set was generated using the parameters $n = 10 \times 10 = 100$, $p_{TA} = 1$, $\alpha = 0.5$, $\lambda = 15$, $\xi_1 = 1$, $\xi_2 = 10$, $L = 50$, $N = 200$ and $p_i(0) = 0.05$, $i = 1, 2, \ldots, n$. The size of the training data set was 10 000.

The input parameters in the simulation used the same values of $n = 10 \times 10 = 100$, $p_{TA} = 1$, $\alpha = 0.5$, and $\lambda = 15$, and we also specified $\gamma = 0.9$ and P^* with $\varepsilon = 0.05$, $r = 5$, $\eta = 100{,}000$, $\delta = 0.99$ and $\rho = 0.001$. The number of epochs in the simulation was $M = 30$.

The initial weights w were generated by the corresponding procedures of the PyTorch library. The optimizer used in the simulation was the ADAM optimizer from the PyTorch library.

The average time required for training the prediction network was approximately 10 minutes (on the PC described above), which is a practically reasonable time for an offline procedure. Note that after offline training, online decision-making is conducted directly by immediate choice without additional calculations.

The results of the simulations are shown in Figure 6.6. The presented graph was obtained by averaging the temporal difference learning errors over 10,000 pairs in the data set.

Temporal difference learning error decreases both in the training stage and in the validation stage of the learning process, and the smoothed graphs for both stages are exponential graphs with

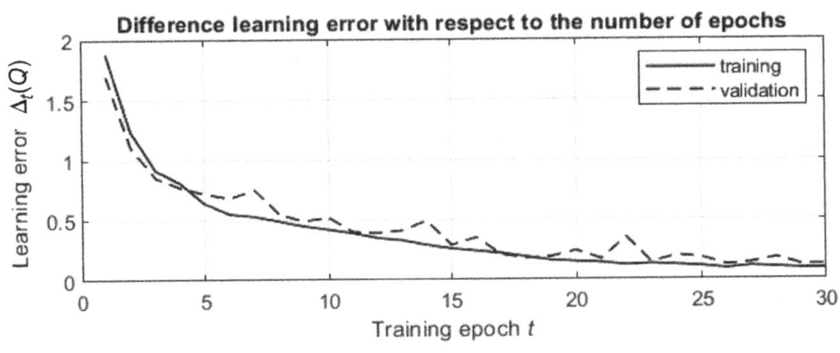

Figure 6.6 Change in the temporal difference learning error with respect to the number of training epochs. The solid line is associated with the training stage, and the dashed line is associated with the validation stage.

similar rates of decrease. This validates the effectiveness of the learning process and shows that progress in network training leads to better processing of previously unknown data from the validation data set.

6.3.2 Detection by the Q-max and SPL Algorithms

In the next simulations, we considered the detection process with the proposed Q-max and SPL algorithms and compared both algorithms with random detection, which provides the lower bound of the cumulative reward (for the Q-max algorithm) and payoff (for the SPL algorithm).

Both algorithms used the same neural network as above, and the random detection process was initialized with the same parameters as above. However, for better comparison, in the simulations of both algorithms and of random detection, we used the same number of targets $\xi = 2$, which were located at the points (5, 0) and (0, 9), and the initial position of the agent was $c(0) = (9, 4)$. By choosing these positions of the targets and the agent, it is easy to demonstrate (i) the difference between the search processes (in which the agent first moves to the closest target and then to the distant target) and the detection process (in which the agent moves to the point that provides the best observation of both targets) and (ii) the motion of the agent over the domain to maximize the immediate reward or minimize the immediate payoff.

The results of the simulations are shown in Figure 6.7. Figure 6.7a shows the discounted cumulative reward for the Q-max algorithm compared to that of the random detection process, and Figure 6.7b shows similar graphs for the SPL algorithm and the random detection process.

The detection by the proposed algorithms is much better than the detection by the random procedure. Namely, the Q-max algorithm results in 20.5 units of discounted cumulative reward, while the random procedure achieves only 13.4 units of discounted reward in the same number of steps. In other words, the Q-max algorithm is nearly 1.5 times more effective than the random procedure. Similarly, while the random procedure requires 40 steps to detect the targets, the SPL algorithm requires only 20 steps, which means that the SPL algorithm is 50% better than the random procedure.

From these comparisons, it follows that the Q-max and SPL algorithms outperform the random procedure in terms of both the informational reward and the agent's path length. However, as follows from the next simulations, the numbers of agent actions up to termination in the Q-max and SPL algorithms are statistically equal, allowing either algorithm to be applied with respect to the considered practical task.

6.3.3 Comparison Between Q-max and SPL Algorithms and One-Step Heuristics

The third set of simulations included comparisons of the Q-max and SPL algorithms with the presented above heuristic methods that implement one-step optimization.

Recall that the simplest algorithm is based on the expected information gain, which is an immediate expected information reward

$$EIG(\mathbb{a}, t) = R(\mathbb{a}, t) = D_{\mathrm{KL}}(P_{\mathbb{a}}(t+1) \| P(t))$$

as mentioned above, here, $P_{\mathbb{a}}(t+1)$ stands for the probability map that is expected to represent the targets' locations at time $t+1$ given that at time t, the agent chooses action $\mathbb{a}(t) \in \mathbb{A}$ and $P(t)$ is the current probability map. Then, the next action is chosen as

$$\mathbb{a}(t+1) = \arg\max_{\mathbb{a} \in \mathbb{A}} \mathrm{EIG}(\mathbb{a}, t)$$

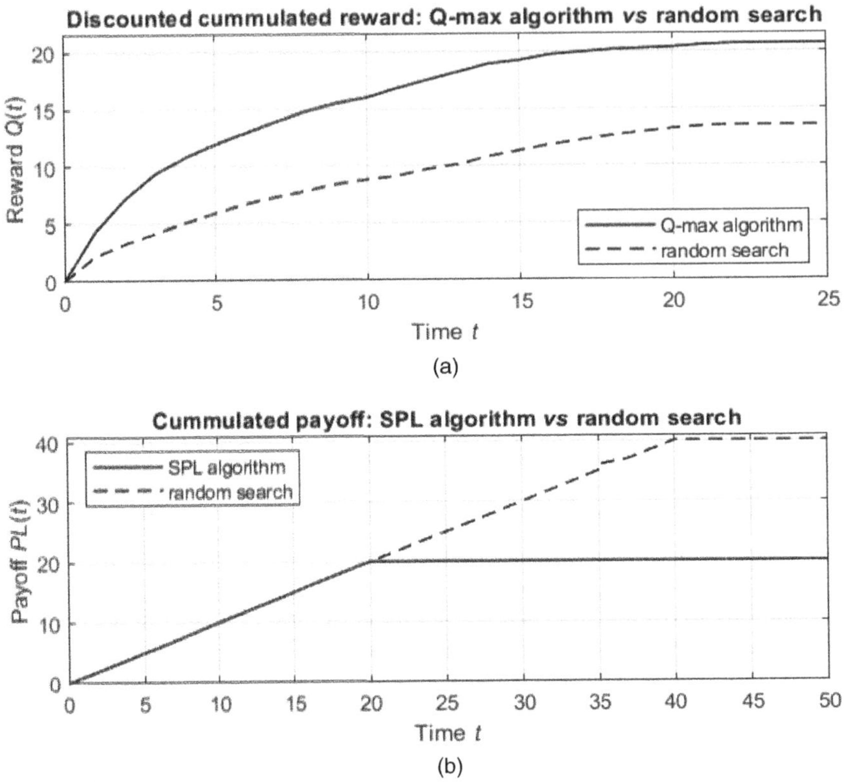

Figure 6.7 Discounted cumulative reward of detection by the Q-max algorithm (a) and cumulative payoff of detection by the SPL algorithm (b) compared with the results obtained by the random detection procedure. The solid line in both figures is associated with the Q-max and SPL algorithms, and the dashed line is associated with the random choice of actions.

More sophisticated algorithm addresses the center of view, which is defined as the cell in which the agent can obtain the maximal expected information gain

$$\mathrm{COV}(t) = \arg\max_{c \in C} D_{\mathrm{KL}}(P_c(t+1) \| P(t))$$

where $P_c(t+1)$ is a probability map that is expected to represent the targets' locations at time $t+1$ when the agent is located in cell c. Then, the next action is chosen as

$$\mathbb{a}(t+1) = \arg\min_{\mathbb{a} \in \mathbb{A}} d(\mathrm{COV}(t), \mathbb{a}(c(t)))$$

where $d(\mathrm{COV}(t), \mathbb{a}(c(t)))$ is the distance between the center of view $\mathrm{COV}(t)$ and cell $\mathbb{a}(c(t))$, to which the agent moves from its current location $c(t)$ when it executes action $\mathbb{a}$.

Note that in contrast to the next location $c(t+1)$, which is one of the neighboring cells of the current agent location $c(t)$, the center of view $\mathrm{COV}(t)$ is a cell chosen from among all n cells of the domain.

Finally, in the third algorithm, the next action is chosen as

$$\mathbb{a}(t+1) = \arg\min_{\mathbb{a} \in \mathbb{A}} d(\mathrm{COG}(t), \mathbb{a}(c(t)))$$

where $\mathrm{COG}(t)$ stands for the "center of gravity," which is the first moment of the probability map $P(t)$, and the remaining terms have the same meanings as above.

Table 6.1 Detection times and accumulated information gain in the detection of $l = 5$ targets by $m = 2$ agents with the sensor sensitivities $\lambda_{1k} = 12$ and $\lambda_{2k} = 8$, $k = 1, 2$.

Detection algorithm	Number of actions up to detection of the first target	Number of actions up to detection of the second target	Discounted cumulative information gain
Random	25	45	13.4
EIG	17	27	17.1
COV	17	24	17.5
COG	18	29	16.1
Q-max	15	21	20.5
SPL	14	21	20.1

The Q-max and SPL algorithms used the same neural network as above and were initialized with the same parameters. As mentioned above, for all the algorithms, the agent started in the initial position $c(0) = (9, 4)$ and moved over the domain with the aim of detecting $\xi = 2$ targets.

The first simulations addressed the detection of static targets, which, as above, were located at points (5, 0) and (0, 9).

The results of the detection by different algorithms are summarized in Table 6.1. The results represent the averages over 30 trials for each algorithm.

The table shows that the proposed Q-max and SPL algorithms outperform previously developed methods in terms of both the number of agent actions and the value of the discounted cumulative information gain.

The results of the simulations over time are shown in Figure 6.8. Figure 6.8a shows the discounted cumulative reward for the Q-max algorithm compared with the COV algorithm (the best heuristic algorithm) and the random detection process, and Figure 6.8b shows similar graphs for the SPL algorithm compared to the COV algorithm and the random detection process.

Detection by the SPL algorithms is better than the detection by the COV algorithm. Namely, the Q-max algorithm results in 20.5 units of discounted cumulative reward, while the COV algorithm obtains 17.5 units of discounted reward in the same number of steps. In other words, the Q-max algorithm is nearly 1.15 times more effective than the COV algorithm. Similarly, while the COV algorithm requires 25 steps to detect the targets, the SPL algorithm requires only 20 steps, which means that the SPL algorithm is 25% better than the COV algorithm.

The second simulation addressed the detection of moving targets, which started in the initial positions (5, 0) and (0, 9). Regarding the targets' motion, it is assumed that both of them, at each time $t = 1, 2, \ldots$, can apply one of the possible actions from the set $\mathbb{A} = \{\uparrow, \nearrow, \rightarrow, \searrow, \downarrow, \swarrow, \leftarrow, \nwarrow, \odot\}$ so that the probability of the action $\odot$ is $\Pr\{\mathbb{a}(t) = \odot\} = 0.9$ and the probability of each other action $\mathbb{a} \in \mathbb{A} \backslash \odot$ is $(1 - 0.9)/8 = 0.0125$.

Results of detection by different algorithms (averaged over 30 trials for each algorithm) are summarized in Table 6.2.

In detection of moving targets, the Q-max and SPL algorithms also outperform previously developed methods in terms of both the number of agent actions and the value of the discounted cumulative information gain.

Note that the simulation was conducted for targets with a clear motion pattern, where the probabilities of their actions represent a slow random motion of the targets near their initial locations.

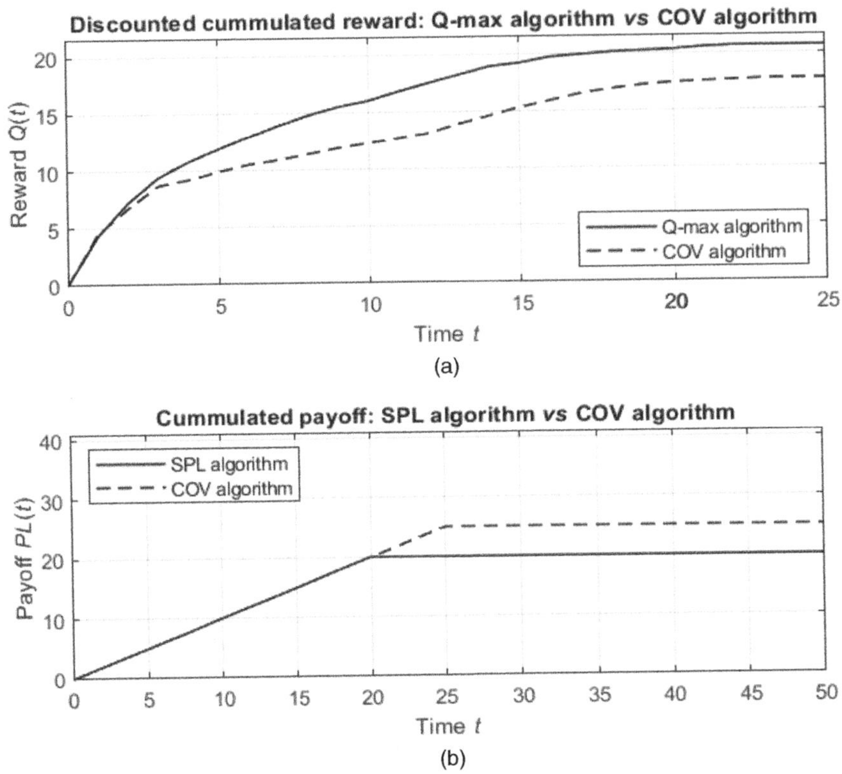

Figure 6.8 The cumulative reward of detection by the Q-max algorithm for static targets (a) and cumulative payoff of detection by the SPL algorithm for static targets (b) compared with the results obtained by the COV algorithm.

Table 6.2 The number of agent actions and the discounted cumulative information gain in detecting two moving targets for the false-alarm rate $\alpha = 0.5$.

Detection algorithm	Number of actions up to detection of the first target	Number of actions up to detection of the second target	Discounted cumulative information gain
Random	72	105	21.8
EIG	50	65	27.1
COV	49	62	28.7
COG	55	67	26.2
Q-max	32	45	33.2
SPL	31	43	32.1

Another possible reasonable motion pattern is motion with a strong drift in a certain direction, which results in a similar ratio between the numbers of actions and the discounted cumulative information gains to that presented in Table 6.2.

In contrast, if the random motion of the targets is a random walk with equal probabilities $\Pr\{\mathbb{a}(t) = \mathbb{a}\} = 1/9$ for all actions $\mathbb{a} \in \mathbb{A}$, then the training becomes meaningless since, both with and without training, the agent needs to detect randomly moving targets.

Table 6.3 The number of agent actions and the discounted cumulative information gain in detecting two moving targets for the false-alarm rate $\alpha = 0.5$.

Sensor sensitivity	Algorithm	False-alarm rate		
		$\alpha = 0.25$	$\alpha = 0.5$	$\alpha = 0.75$
$\lambda = 15$	COV	14	25	63
	SPL/Q-max (average)	13	20	45
$\lambda = 5$	COV	64	95	242
	SPL/Q-max (average)	44	54	63

The other results obtained for the Q-max/SPL algorithms also indicated better performance by these algorithms compared with that of the heuristic algorithms. The algorithms were compared with the best heuristic COV algorithm. The results of the trials for different values of the false-alarm rate α and of the sensor sensitivity λ are summarized in Table 6.3.

For all values of the false-alarm rate and the sensor sensitivity, the Q-max and SPL algorithms strongly outperform the best heuristic COV algorithm.

To emphasize the difference in the detection time between the SPL and Q-max algorithms and the heuristic COV algorithm, the data shown in the table are depicted in Figure 6.9.

As expected, the Q-max and SPL learning algorithms demonstrate better performance than the heuristic COV algorithms without learning, and the difference between the algorithms increases as the false-alarm rate α increases, and the sensor sensitivity λ decreases. For example, if $\lambda = 15$ and $\alpha = 0.25$, then the improvement in the number of actions is 10%, while if $\lambda = 5$ and $\alpha = 0.75$, then the improvement is significantly stronger at 75%.

In other words, computationally inexpensive heuristic algorithms provide effective results in searches with accurate sensors and a low rate of false alarms. However, in searches with less precise sensors or a high rate of false-positive errors, the heuristic algorithms are less effective, and the Q-max and SPL learning algorithms should be applied.

6.3.4 Comparison Between SPL Algorithm and Optimal Solution

The considered approach was compared with the known dynamic programming techniques implemented in search algorithms for moving targets (for overview, see e.g. [15, 36]). These algorithms

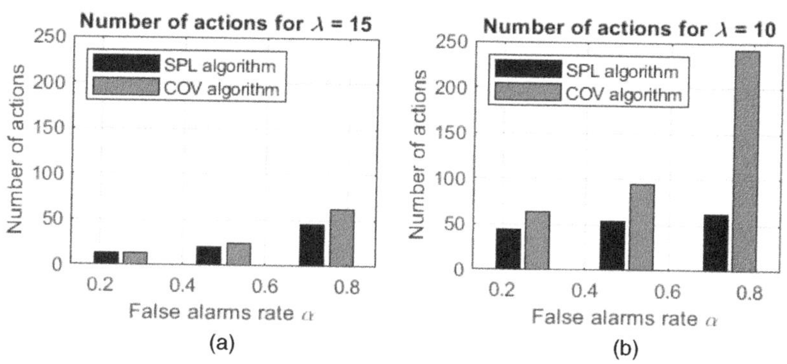

Figure 6.9 Number of agent actions in detecting two static targets with the SPL/Q-max algorithms (black bars) and the COV algorithm (gray bars): (a) $\lambda = 15$ and (b) $\lambda = 10$.

directly address the optimal trajectory of the agent and result in an optimal path. So, in the simulation we considered the SPL algorithm, which uses the same criteria as the known algorithms.

Comparisons were conducted as follows. The algorithms were trialed over the same domain with a definite number n of cells, and the goal was to reach the maximal probability $P^* = 0.95$ of detecting the target. When this probability was reached, the trial was terminated, and the number of agent actions was recorded.

Since the known algorithms implement dynamic programming optimization over possible agent trajectories, their computational complexity is high, and for the considered task, it is $\mathcal{O}(n \cdot 9^t)$, where n is the number of cells and t is the number of actions.

Therefore, to finish the simulations in a reasonable time (120 minutes for each trial), the algorithms were trialed on a very small case with $n = 10 \times 10 = 100$ cells. Note that in the original simulations, these algorithms were trialed in smaller domains. If the desired probability $P^* = 0.95$ of detecting the targets was not reached in 120 minutes, the algorithms were terminated.

In all trials, the known dynamic programming algorithms planned $t = 7$ agent actions in 120 minutes, while the SPL algorithm, during the same time of 120 minutes, planned significantly more actions and reached at least the desired probability $P^* = 0.95$ of detecting the targets. The results of the comparison between the SPL algorithm and the known dynamic programming algorithms that provide optimal solutions are summarized in Table 6.4.

Until termination at 120 minutes, the SPL algorithm plans more agent actions and results in higher detection probabilities than the DP algorithm for both values of sensor sensitivity λ and all values of the false-alarm rate α. For example, the dependence of the detection probabilities on the run time for sensor sensitivity $\lambda = 15$ and false-alarm rate $\alpha = 0.25$ is depicted in Figure 6.10.

Table 6.4 The number of planned agent actions in detecting two static targets by the SPL algorithm and dynamic programming (DP) algorithm for different values of the false-alarm rate α and sensor sensitivity λ.

			False-alarm rate				
Sensor sensitivity	**Algorithm**	**Characteristic**	$\alpha = 0$	$\alpha = 0.05$	$\alpha = 0.1$	$\alpha = 0.25$	$\alpha = 0.5$
$\lambda = 15$	DP	Run time	0.4 s	1 min	120 min	120 min	120 min
		Number of planned actions	3	5	7	7	7
		Detection probabilities p_1 and p_2	1.0 1.0	1.0 0.99	0.99 0.96	0.90 0.84	0.84 0.68
	SPL	Run time	0.4 s	1 min	120 min	120 min	120 min
		Number of planned actions	3	5	7	13	20
		Detection probabilities p_1 and p_2	1.0 1.0	1.0 0.99	0.99 0.96	0.99 0.95	0.99 0.95
$\lambda = 10$	DP	Run time	1 min	120 min	120 min	120 min	120 min
		Number of planned actions	5	7	7	7	7
		Detection probabilities p_1 and p_2	1.0 1.0	0.96 0.95	0.90 0.85	0.85 0.65	0.71 0.43
	SPL	Run time	1 min	120 min	120 min	120 min	120 min
		Number of planned actions	5	7	15	21	32
		Detection probabilities p_1 and p_2	1.0 1.0	0.96 0.95	0.97 0.95	0.98 0.95	0.99 0.95

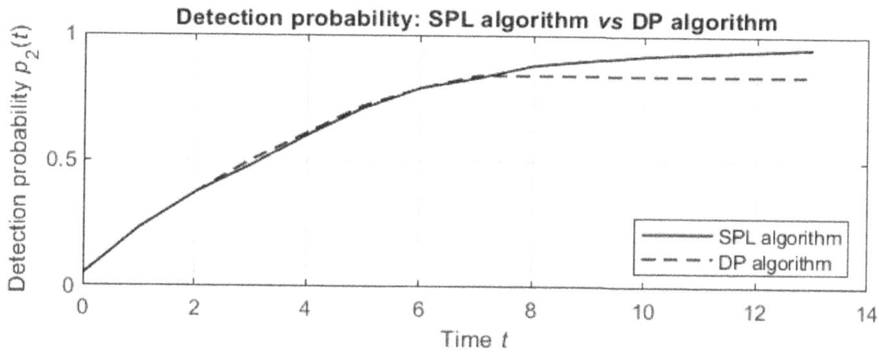

Figure 6.10 Dependence of the detection probabilities on the number of planned actions for the SPL algorithm (solid line) and DP algorithm (dotted line); the sensor sensitivity is $\lambda = 15$, the false-alarm rate is $\alpha = 0.25$, and the termination time is $t = 120$ minutes.

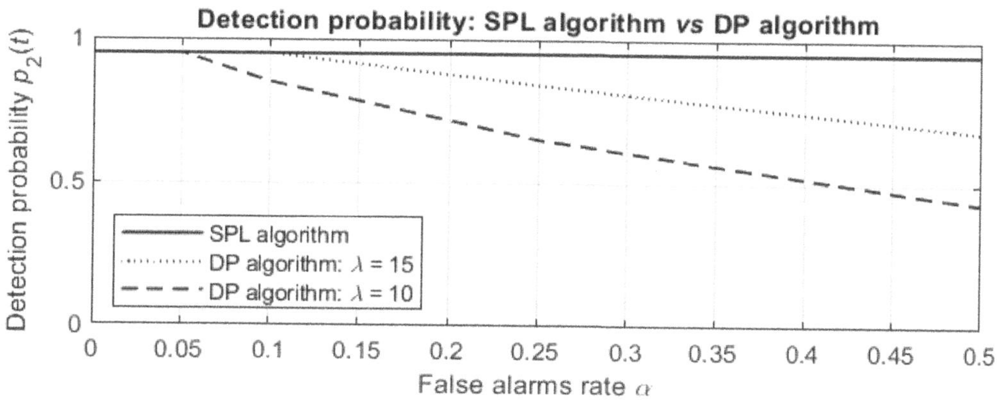

Figure 6.11 Dependence of the detection probabilities on the false-alarm rate α for sensor sensitivities $\lambda = 15$ (dotted line) and $\lambda = 10$ (dashed line). The probability 0.95 for the SPL algorithm, and all values of α is depicted by the solid line. The termination time is 120 minutes.

For the first 7 actions, the detection probabilities of both algorithms increase similarly. Then, the DP algorithm does not plan additional actions in 120 minutes, while the SPL algorithm results in more planned actions, and the detection probabilities for these actions continue increasing until termination after 13 planned actions.

Finally, the dependence of the detection probabilities on the false-alarm rate α at termination after 120 minutes is depicted in Figure 6.11.

For a low false-alarm rate α, the SPL algorithm results in the same detection probabilities as the optimal DP algorithms. However, for a higher false-alarm rate α, the detection probabilities obtained by the DP algorithms significantly decrease (to 0.68 and 0.43 for $\lambda = 15$ and $\lambda = 10$, respectively), while the probability obtained by the SPL algorithm is 0.95 for any false-alarm rate and both sensor sensitivities.

6.3.5 Run Times and Mean-Squared Errors for Different Sizes of Data Sets

Finally, let us consider dependence of the run time and mean-squared error on the size of the data set. Results of these simulations are summarized in Table 6.5.

Table 6.5 Run times and temporal difference errors with respect to the size of the data set.

Domain size $n_x \times n_y$	Number of nonzero weights in the network	Size of the data set	Run time for one epoch (minutes)	Mean Squared Error[a)]
10×10	42,009	5,000	4	0.13
		10,000	8	0.12
20×20	648,009	5,000	7	0.15
		10,000	14	0.13
40×40	10,272,009	5,000	10	0.18
		10,000	20	0.15

a) The error was calculated over the temporal difference errors at the validation stage at epoch $t = 30$.

While the size of the domain and the number of links in the network increase, the mean squared error increases very slowly and remains small.

An exponential increase of the domain size leads to linear increase of the run time, and the computations require a reasonable time even on the previously described PC.

However, for realistic engineering and industrial problems with larger domains, it is reasonable to use computation systems with greater GPU power.

6.4 Conclusions

In the chapter, we considered an algorithm for navigation of search agent detecting static and moving hidden targets in the presence of false-positive and false-negative errors. In contrast to the previously considered procedures that follow an immediate one-step decision-making process, the proposed algorithm implements the deep Q-learning approach and neural network techniques.

The algorithm is implemented in two versions: a procedure that maximizes the cumulative discounted expected information gain over the domain (Q-max algorithm) and a procedure that minimizes the expected path length of the agent in detecting all the targets (SPL algorithm).

Formally, the first procedure is an extension of previously developed techniques based on the expected information gain calculated over the local neighborhood of the agent, and the second is a direct application of Q-learning techniques to the required value of the agent's path length.

Simulations show that after offline training of the neural network using the generated data set, the algorithm provides solutions that outperform the results obtained by the previously developed procedures, both in terms of the cumulative information gain and in terms of the agent's path length.

The expected number of actions obtained by the Q-max algorithm by maximizing the cumulative discounted expected information gain is statistically equal to the number of actions obtained by the SPL algorithm by minimizing the expected path length.

This equivalence follows directly from the nature of the problem: in terms of information, detection of the targets means accumulating as much information as possible about the targets' locations, and in terms of the path length, the detection of the targets means making as few movements as possible to specify the exact target locations.

7

Search with Deep Q-learning: Multiple Agents

In this chapter, we continue consideration of deep Q-learning in search and detection problems and present the method of search for multiple static and moving targets by a group of search agents. Following conventional formulations, we consider a search with constrained paths and simulate the detection uncertainty using intermittently emitting targets.

The considered solution includes two algorithms: the distributed expected information gain (DEIG) algorithm, which implements dynamic Voronoi partitioning of the search space and plans the paths by maximizing the expected one-step look-ahead information per region, and the collective Q-max (CQM) algorithm, which finds the shortest paths of the agents in the group by maximizing the discounted cumulative information about the targets' locations using deep Q-learning techniques.

7.1 Problem Formulation and Notation

In the chapter we consider two algorithms of search for mobile targets by a team of search agents such that detection can include the errors of the first and the second types. The first algorithm of quick online search implements dynamic Voronoi partitioning of the domain and plans the search paths by maximizing an expected one-step information gain in each region.

Following the algorithm, at each moment, the domain is divided into Voronoi regions with respect to the current probability map of the domain, and each agent searches in its region independently from the other agents. During the search, location probabilities of the targets over the search domain change, causing the Voronoi regions to change, and thus, directly affecting the agents' search movements and plan over the updated regions.

This version of the algorithm is called the DEIG algorithm in contrast to the previously considered centralized EIG algorithm (see Section 7.4) that did not use the Voronoi diagram. A clear benefit of this heuristic is its simplicity and low complexity, which allows online implementation over simple search agents.

The second algorithm finds the shortest paths of the search agents by maximizing the cumulative information about the targets' locations using deep Q-learning techniques. Decision about the next step of the search agent is obtained by a deep Q-learning scheme over a neural network based on agent-by-agent value iteration.

The network receives the agent's location, the current probability map and the networks' parameters of previous agents as input, and outputs the preferred move of the agent. This algorithm maximizes the network Q-value over a group of agents and is called the CQM algorithm.

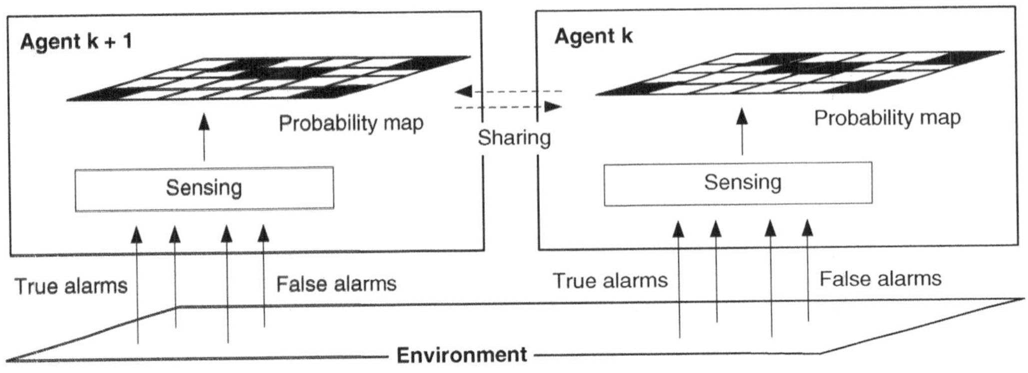

Figure 7.1 Receiving information and updating a shared probability map.

In the chapter, we follow the notation used in Chapter 6. The search agents act on the set $C = \{c_1, c_2, \dots, c_n\}$ of cells, which represents a two-dimensional gridded domain and search for several targets hidden in or moving over the set C. The targets are not aware of the searchers' activity.

The number of searchers is η, $1 \le \eta \le n-1$ (in practice, $\eta \ll n$) and the number of targets is ξ, $\xi \le n-1$. Each search agent and each target can occupy only one cell.

As above, it is assumed that each search agent is equipped with several sensors that can detect the targets in different directions and distances. Sensitivity of the sensors is denoted by λ with appropriate indices.

Similar to Chapter 6, detection probabilities are defined using the Koopman random search formula, the errors of the first and the second types are represented by true and false alarms $\tilde{a}(c,t)$ broadcasted by the cells $c_i \in C$, $i = 1, 2, \dots, n$, and probabilities $p_i(t)$ of targets' location in the cells c_i at time t form a probability map $P(t) = \{p_1(t), p_2(t), \dots, p_n(t)\}$.

Each searcher creates and updates its probability map with respect to the perceived information and shares the map with the other agents. As a result, all agents have access to the same information about the search process and make decisions with respect to this information.

The process of receiving signals and updating the probability map is illustrated in Figure 7.1.

In the case of static targets, the targets' location probabilities $p_i(t)$, $i = 1, 2, \dots, n$, depend only on the agents' positions at time t and their movements, and in the case of moving targets, these probabilities are defined both by the targets' and by the agents' motion.

The goal of the search agents is to detect all ξ targets in a minimal time.

7.2 Cooperative Detection Using Voronoi Regions and Deep Q-Learning

The considered detection process is as follows: being at time t in cell $c(t)$ each agent k obtains the probability map $P(t)$, receives the alarms $\tilde{a}(c,t)$ from the available cells and decides to which cell $c(t+1)$ it should move.

Accordingly, the main challenge for an efficient cooperative search process is to divide the domain into several regions and to plan the search for each agent such that overall detection time will reach its minimum.

Below, we present two algorithms that address the challenge: the first considers the Voronoi regions of the agents and plans the agents' motion in their regions, and the second uses the shared probability map and implements deep Q-learning techniques to control the agents' search.

7.2.1 Agents' Actions and Decision-making

Actions and decision-making process of each agent are defined similarly to the actions and decision-making in the search by single agent considered in Section 6.2.1. For convenience we recall these definitions.

Consider the kth search agent which at time t in located in the cell $c(t) \in C$. As above, we assume that the agent can choose an action

$$\mathbb{a}^k(t) \in \mathbb{A} = \{\uparrow, \nearrow, \rightarrow, \searrow, \downarrow, \swarrow, \leftarrow, \nwarrow, \odot\}$$

The probability map of the kth agent at time $t+1$ is denoted by $P^k_{\mathbb{a}}(t+1)$, given that at time t, the kth agent will choose action $\mathbb{a}^k(t)$. Then, given the action $\mathbb{a}^k(t)$, the immediate expected informational reward of the kth agent is given by the Kullback–Leibler distance

$$R^k_{\mathbb{a}}(t) = D_{\mathrm{KL}}\left(P^k_{\mathbb{a}}(t+1) \| P(t)\right)$$

between the expected agent's map $P^k_{\mathbb{a}}(t+1)$ and the current probability map $P(t)$. This informational reward $R^k(t)$ forms a basis for making decisions about the agents' next steps.

In search by a single agent or by several agents making independent decisions over a shared probability map, the choice of the next action is governed by the simple rule

$$\mathbb{a}^k(t) = \underset{\mathbb{a} \in \mathbb{A}}{\operatorname{argmax}}\, R^k_{\mathbb{a}}(t)$$

applied by each search agent $k = 1, 2, \ldots, \eta$. This rule represents the agents' immediate one-step reaction to the changes in the probability maps and states of the targets.

Below we extend this rule to the search of multiple targets by a group of several interacting agents. The first approach is based on the Voronoi diagrams, and the second uses deep-learning techniques.

7.2.2 Reactive Decision-making in Voronoi Regions: A Distributed *EIG* Algorithm

Let $C = \{c_1, c_2, \ldots, c_n\}$ be a two-dimensional domain and $P(t) = \{p_1(t), p_2(t), \ldots, p_n(t)\}$ be a probability map at time t. The map $P(t)$ is shared among all η agents, $\eta \le n-1$ and is updated with respect to the information obtained by each kth agent, $k = 1, 2, \ldots, \eta$.

The Voronoi region for each search agent k is defined as follows. Assume that at time t, the kth agent is in the cell $c^k(t)$ and $d(c^k(t), c)$ are the distances between cell $c^k(t)$ and cell $c \in C$ of the domain. Then, the Voronoi region of agent k is a subset $C^k(t) \subset C$

$$C^k(t) = \{c \mid d(c^k(t), c) < d(c^j(t), c),\ j = 1, 2, \ldots \eta,\ j \neq k\},$$

which includes the cells that are closer to the position of the kth agent than to the positions of the other agents. Since the agents change their positions with time, the Voronoi regions of the agents are updated accordingly.

The part of the probability map which corresponds to the Voronoi region $C^k(t) \subset C, k = 1, 2, \ldots, \eta$, is denoted by $P^k(t) \subset P(t)$. The probability maps $P^k(t)$ are updated simultaneously with the updates of the regions $C^k(t)$, and the values of the probabilities $p(t) \subset P^k(t)$ are updated according to the detection results. The set of Voronoi regions $C^k(t)$ is called the Voronoi diagram and denoted by $\mathfrak{C}$, and the set of probability maps $P^k(t)$ is called the probability atlas and is denoted by $\mathfrak{P}$.

Based on the Voronoi regions, the detection process is conducted in several steps. Given positions $c^k(t)$ of agents $k = 1, 2, \ldots, \eta$, domain C is divided into Voronoi regions $C^k(t)$. Each kth agent focuses on the corresponding part $P^k(t)$ of the probability map $P(t)$ and by the rule makes an independent decision regarding its next movement. When all η agents make their decisions, move to the next positions $c^k(t+1)$ and observe the sensors' output, they update the location probabilities to $p(t+1)$ to form an updated probability map $P(t+1)$, and the process continues.

Formally, the distributed EIG algorithm based on the Voronoi regions is outlined as follows.

Algorithm 7.1 Cooperative detection with reactive decision-making: distributed EIG algorithm

---

Input: Set $C = \{c_1, c_2, \dots, c_n\}$,
number of search agents η,
initial agents' locations $c^1(0), c^2(0), \dots, c^\eta(0)$
set $\mathbb{A} = \{\uparrow, \nearrow, \rightarrow, \searrow, \downarrow, \swarrow, \leftarrow, \nwarrow, \odot\}$ of possible movements,
probability p_{TA} of true alarms,
rate α of false alarms and probability $p_{\text{FA}} = \alpha p_{\text{TA}}$,
sensor sensitivity λ,
initial probability map $P(0) = \{p_1(0), p_2(0), \dots, p_n(0)\}$,
number of targets ξ.

Output: targets' locations $\hat{c}^1(T), \hat{c}^2(T), \dots, \hat{c}^\xi(T)$ at a termination time T.

---

1. Start with $t = 0$, initial agent positions $c^1(t), c^2(t), \dots, c^\eta(t)$ and initial probability map $P(t) = \{p_1(t), p_2(t), \dots, p_n(t)\}$.
2. Create the Voronoi diagram $\mathfrak{C}(t) = \{C^1(t), C^2(t), \dots, C^\eta(t)\}$, $C^k(t) \subset C$, $k = 1, 2, \dots, \eta$.
3. Create the probability atlas $\mathfrak{P}(t) = \{P^1(t), P^2(t), \dots, P^\eta(t)\}$, $P^k(t) \subset P(t)$, $k = 1, 2, \dots, \eta$.

<u>Decision-making</u>

4. For each agent $k = 1, 2, \dots, \eta$ do:
5. Choose action $\mathbb{a}^k(t) = \underset{\mathbb{a} \in \mathbb{A}}{\operatorname{argmax}}\, R^k_{\mathbb{a}}(t)$, where $R^k_{\mathbb{a}}(t) = D_{\text{KL}}\left(P^k_{\mathbb{a}}(t+1) \middle\| P^k(t)\right)$.
6. End for

<u>Acting</u>

7. For each agent $k = 1, 2, \dots, \eta$ do:
8. Apply action $\mathbb{a}^k(t)$: move to new position $c^k(t+1)$.
9. End for

<u>Updating</u>

10. Set $t = t + 1$.
11. For each agent $k = 1, 2, \dots, \eta$ do:
12. Screen the domain C with respect to the sensor's abilities.
13. Update the probability map $P(t)$ to $P(t+1)$.
14. End for
15. If all ξ targets are detected, then
16. Set $T = t$ and terminate (go to line 20).
17. Else
18. Continue with line 10.
19. End if.
20. Return targets' locations $\hat{c}^1(T), \hat{c}^2(T), \dots, \hat{c}^\xi(T)$.

In this algorithm, it is assumed that the number ξ of targets is known, and this number is used to terminate the process. If the number of targets is unknown, then to define termination of the algorithm, one can use certain measures over the probability map P^*, such as entropy of the location probability, which represent general knowledge about the targets' locations. In this case, the condition in line 16 is substituted or completed by the condition concerning the equivalence of the current probability map $P(t)$ and the objective map P^*.

In the simulations below, we assumed that $P^* = (p_1^*, p_2^*, \dots, p_n^*)$ is constant with $p_i^* = 0.95$, $i = 1, 2, \dots, n$, and used this map as a termination condition. In the search with Q-deep learning the objective probability map P^* is used both for learning and for termination; below we consider this scenario in detail.

Note that in real-world search tasks, the map P^* is not necessarily known by the search agents and they can continue their activity, but this map is necessary for control of the algorithm.

Actions of the distributed EIG Algorithm 7.1 are illustrated in Figure 7.2, which shows four detection stages of $\xi = 30$ static targets by $\eta = 6$ agents from a stating time $t = 0$ until $t = 90$ when all the targets are detected. The figures on the left side show the positions of the targets and the agents, and the Voronoi regions and the right-side figures show the probability maps. The detected targets are denoted by white squares and the targets for which the detection probability is less than 0.95 are denoted by gray squares with the brightness proportional to the detection probability.

The presented algorithm is a direct extension of a previously developed EIG algorithm (see Chapter 4) that maximizes the expected information gain (EIG) over a domain C. However, the use of a Voronoi diagram $\mathfrak{C}(t)$ and the independent activity of the agents in their Voronoi regions leads to changes in the agents' motion and a serious decrease in the detection time.

7.2.3 Collective Deep Q-Learning Approach

Now assume that, in addition to sharing the probability map, each agent makes its decisions with respect to the decisions made by the other agents. In this setting, a direct solution to the detection problem is computationally hard, and exact activity of the group of agents can be specified only for small domains C. To overcome this problem, we apply agent-by-agent value iteration [2] together with reinforcement and deep learning techniques [9, 44].

The considered collective Q-learning approach extends the Q-max algorithm of search by a single searcher (see Section 6.2.5). In the algorithm, each kth agent, $k = 1, 2, \dots, \eta$, deals with two neural networks, the prediction network and the target or the Q-max network.

The input layer of each of the networks includes $2n$ neurons, where n is the size of the domain. The first chunk of n inputs $(1, 2, \dots n)$ receives a binary vector that represents the agent's position; if the agent is in cell c_j^k, then the jth input of the network is equal to 1, and the other $n - 1$ inputs are equal to 0. For convenience, the cell occupied by the considered agent is denoted by the value 10 instead of 1. The second chunk of n inputs $(n + 1, n + 2, \dots 2n)$ receives the target location probabilities; the $(n + i)$th input receives the target location probability p_i, $i = 1, 2, \dots, n$, as it appears in the probability map P.

The hidden layer of each network is a fully connected linear layer, which consists of $2n$ neurons with the sigmoid activation function

$$f(x) = 1/(1 + e^{-x})$$

that returns the values in the range (0, 1). The other possibility is to use the SoftPlus techniques with the activation function

$$f(x) = \ln(1 + e^x)$$

that returns the values in the range $(0, \infty)$. In the simulations, it was observed that both functions result in similar performances. However, for both functions the observed performance is significantly better than the performance based on the step activation function.

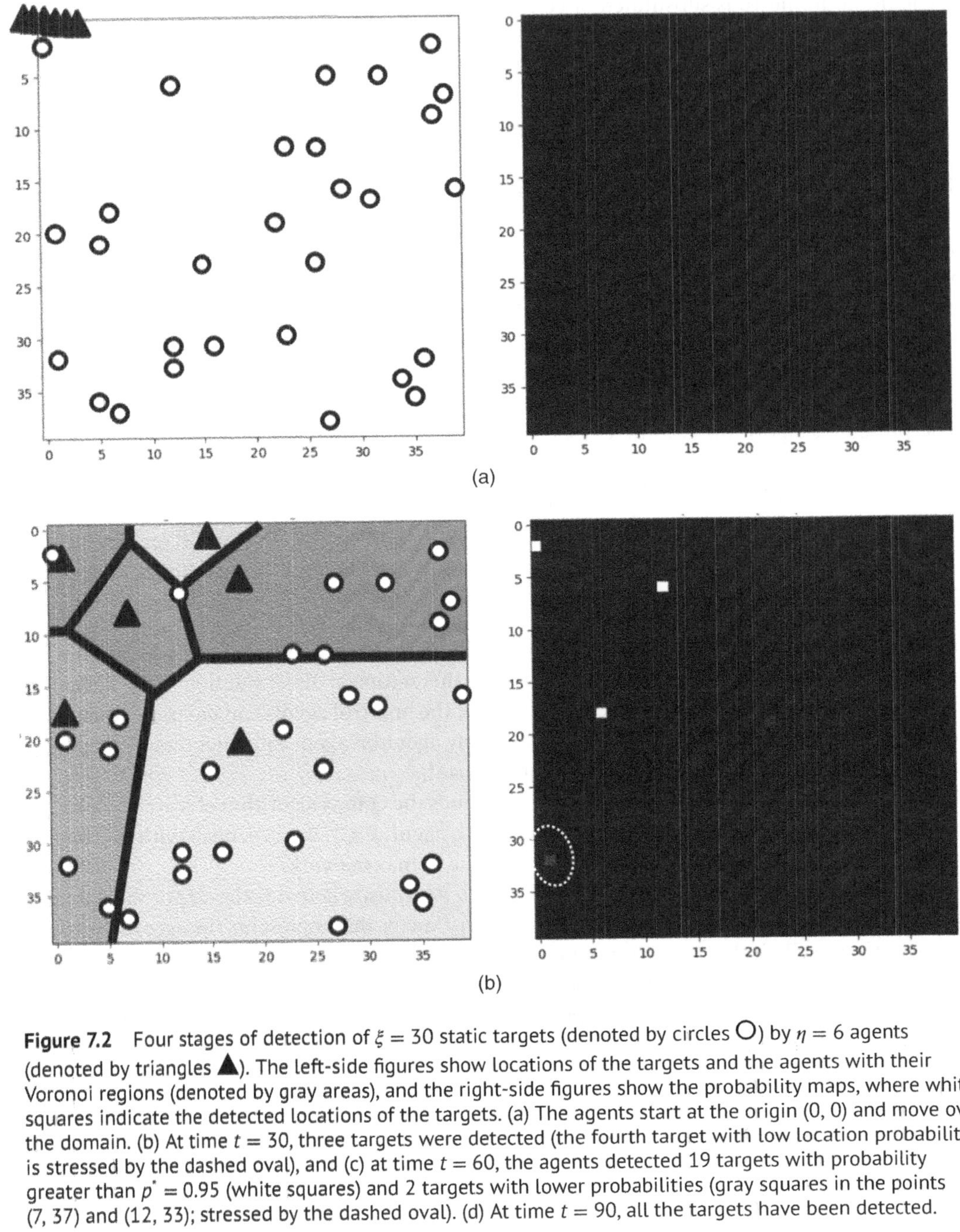

Figure 7.2 Four stages of detection of $\xi = 30$ static targets (denoted by circles O) by $\eta = 6$ agents (denoted by triangles ▲). The left-side figures show locations of the targets and the agents with their Voronoi regions (denoted by gray areas), and the right-side figures show the probability maps, where white squares indicate the detected locations of the targets. (a) The agents start at the origin (0, 0) and move over the domain. (b) At time $t = 30$, three targets were detected (the fourth target with low location probability is stressed by the dashed oval), and (c) at time $t = 60$, the agents detected 19 targets with probability greater than $p^* = 0.95$ (white squares) and 2 targets with lower probabilities (gray squares in the points (7, 37) and (12, 33); stressed by the dashed oval). (d) At time $t = 90$, all the targets have been detected.

The output layer includes 9 neurons with respect to the number $\#\mathbb{A}$ of possible actions. As above, the first output corresponds to the action $\mathbb{a}_1 =$ " $\uparrow$ ", "step forward"; the second output corresponds to the action $\mathbb{a}_2 =$ " $\nearrow$ ", "step right-forward," up to action $\mathbb{a}_9 =$ " $\odot$ ", which implies "stay in the current cell". The value of the ith output is the maximal expected cumulative discounted reward $Q\left(c_j^k, P, \mathbb{a}_i^k\right)$ obtained by the agent if it is in the cell c_j^k, $j = 1, 2, \dots, n$, and, given the probability map $P = (p_1, p_2, \dots, p_n)$, it chooses action $\mathbb{a}_i^k$, $i = 1, 2, \dots, 9$.

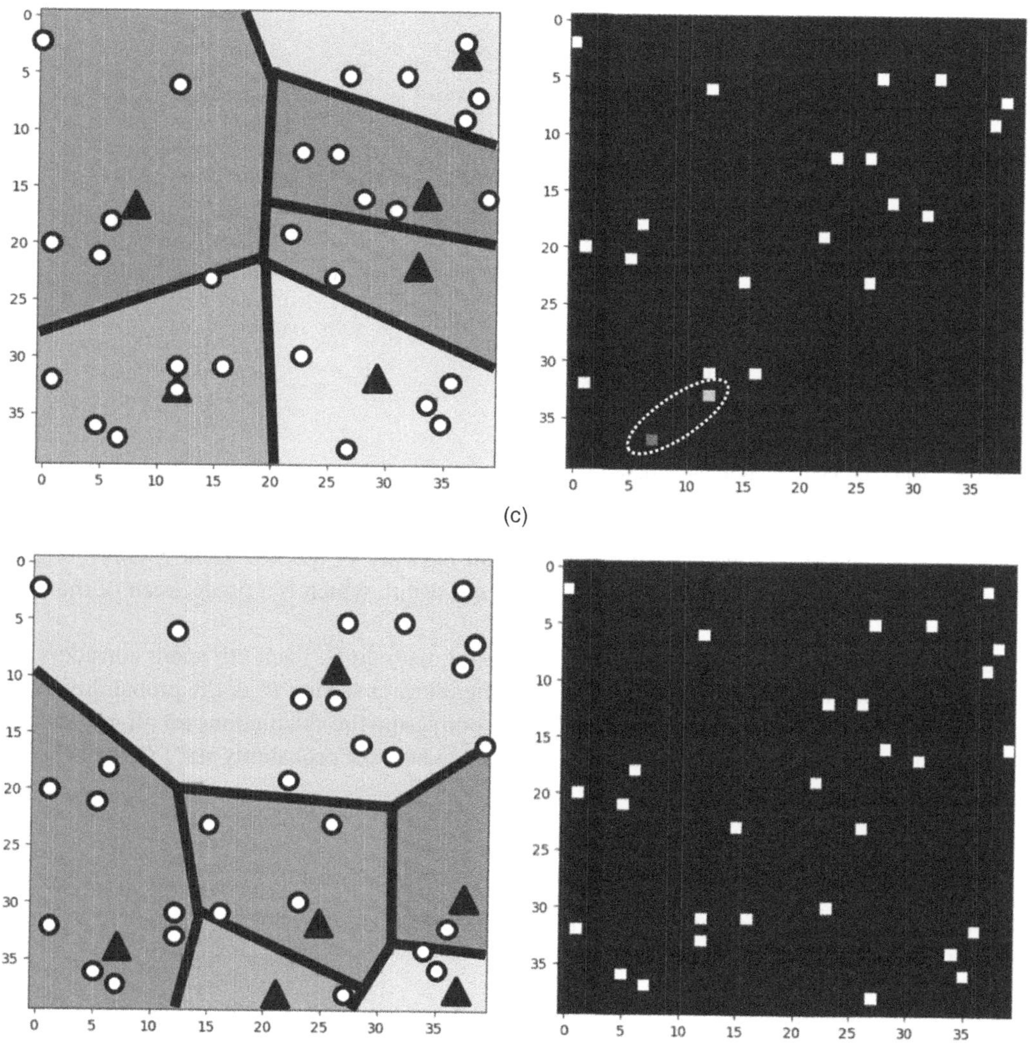

Figure 7.2 *(Continued)*

The scheme of the network of each kth agent is the same as in the Q-max algorithm (see Figure 6.2); for convenience we show this network in Figure 7.3. At the input, the cells occupied by the agents have the values 1, and the cell occupied by the acting agent is denoted by 10.

Interaction between the agents is conducted by two channels: the first is using the shared probability map P, and the second is sharing the updated weights of the links in the prediction network. This means that the $(k+1)$th agent starts training the network with weights specified at the end of the training by the kth agent.

The Bellman equation for calculating the maximal cumulative discounted reward is as follows:

$$Q(c^k(l), P(l), \mathbb{a}^k(l)) = \begin{cases} R^k_{\mathbb{a}}(l) + \gamma \max\limits_{\mathbb{a}\in\mathbb{A}} Q(c^{k+1}(l+1), P(l+1), \mathbb{a}) & k = 1, 2, \dots, \eta - 1, \\ R^1_{\mathbb{a}}(l) + \gamma \max\limits_{\mathbb{a}\in\mathbb{A}} Q(c^1(l+1), P(l+1), \mathbb{a}) & k = \eta, \end{cases}$$

where $l = 1, 2, \dots$ enumerates the steps at the learning stage. This equation forms a basis for updating the weights of the links in the networks.

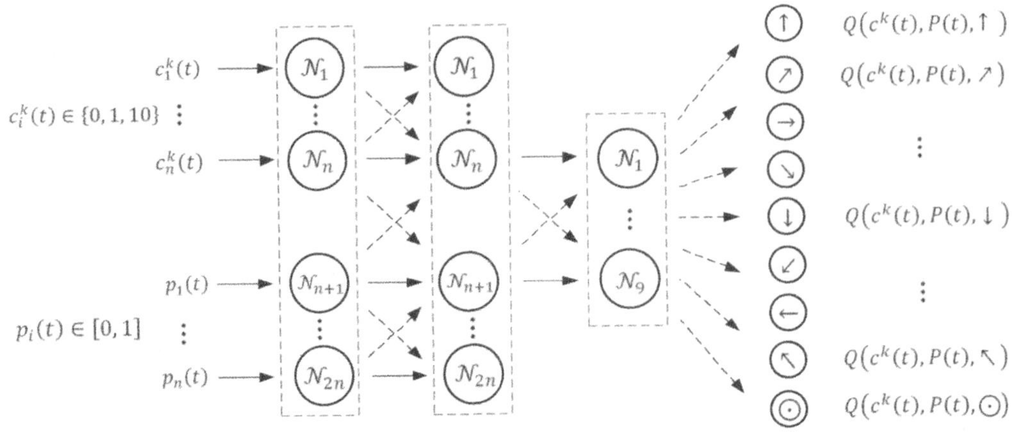

Figure 7.3 The scheme of neural network of the kth agent used in the learning stage of the collective Q-max algorithm.

Thus, in the collective Q-max algorithm, the target network of the kth agent is considered a prediction network for the $(k+1)$th agent up to the ηth agent, which is a predecessor of the first agent.

The agents' rewards are calculated using the Voronoi diagram $\mathfrak{C}$. Each kth agent considers its region $C^k \in \mathfrak{C}$ and calculates the reward $R^k_{\mathbb{a}}$ within the relevant section P^k of the probability map. Updating the Voronoi diagram $\mathfrak{C}$ is conducted after completing the calculations for all η agents.

The SoftMax policy is implemented to select an action, where the probability $p(\mathbb{a}_i \mid Q; \theta)$ of choosing action $\mathbb{a}_i$ is defined as follows:

$$p(\mathbb{a}_i \mid Q; \theta) = \frac{\exp[Q(c, P, \mathbb{a}_i)/\theta]}{\sum_{j=1}^{9} \exp[Q(c, P, \mathbb{a}_j)/\theta]},$$

where $\theta \in [0, +\infty)$ is a parameter that governs the randomness of the choice.

If $\theta \to 0$, then $p(\mathbb{a}_i \mid Q; \theta) \to 1$ for

$$\mathbb{a}_i = \arg\max_{\mathbb{a} \in \mathbb{A}} Q(c, P, \mathbb{a}),$$

and $p(\mathbb{a}_i \mid Q; \theta) \to 0$ for all other actions.

If $\theta \to \infty$, then

$$p(\mathbb{a}_i \mid Q; \theta) \to \frac{1}{9},$$

which corresponds to a randomly chosen action.

In the learning process, it is assumed that the value of the parameter θ decreases in the number of steps l from its maximal value to zero. Then, at the first learning stages, the agent chooses actions randomly, and then along with learning, the agent uses the information about the targets' locations learned by the networks. The first stages with randomly chosen actions are usually interpreted as exploration stages, and the later stages, based on the learned information, are considered exploitation stages.

Training of the networks is conducted using the temporal difference learning error

$$\Delta(Q, l+k-1) = Q^+(c^{k+1}(l+1), P(l+1), \mathbb{a}^{k+1}(l+1); w') - Q(c^k(l), P(l), \mathbb{a}^k(l); w),$$

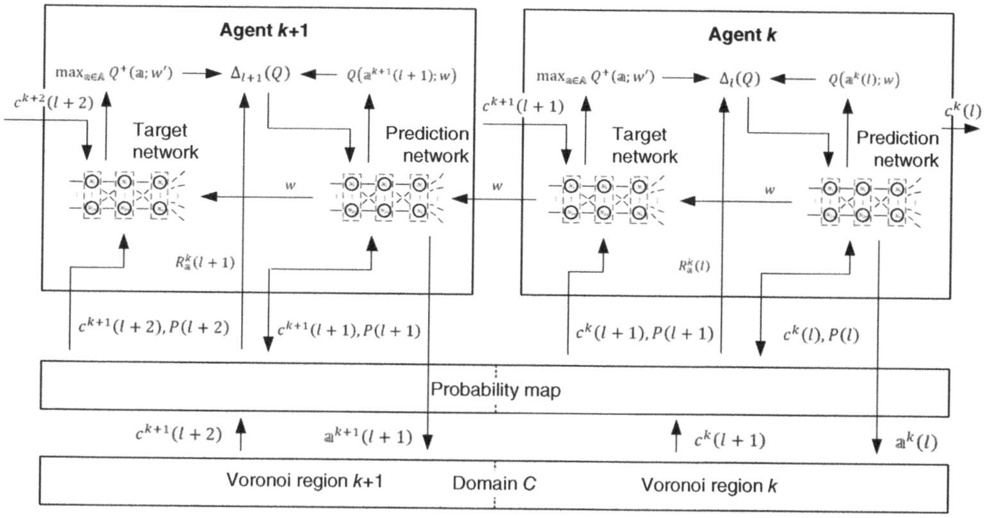

Figure 7.4 The scheme of learning process in the collective Q-max algorithm.

where $Q(c^k(l), P(l), \mathbb{a}^k(l); w)$ denotes the maximal cumulative discounted reward calculated at step $l = 1, 2, \ldots$ by the network with weights w, and $Q^+(c^{k+1}(l+1), P(l+1), \mathbb{a}^{k+1}(l+1); w')$ denotes the expected maximal cumulative discounted reward calculated using the vector w' of the updated weights.

Note that the value of Q^+ is calculated for the next step $l+1$ and for agent $k+1$, which allows sequential computation of the rewards obtained by the agents.

The learning process is illustrated in Figure 7.4.

The prediction network of the kth agent is used for choosing the action and specifying the expected position of this agent at step l, while the target network of the kth agent is used for calculating the reward after selecting and conducting the action that leads to step $l+1$. Then, the cumulative reward calculated by the target network is used to update the prediction network, and its weights are then used to update the weights of the trained network.

Note that together with the position of the kth agent, the target network considers the position of the $(k+1)$th agent. The target network of the kth agent, which at step l was trained with respect to the positions of the kth and $(k+1)$th agent is considered a prediction network of the $(k+1)$th agent at step $l+1$.

We assume that at the learning stage, the search agents share the probability map and that each agent updates the probabilities of the targets' locations over the entire domain. The reward of the kth agent is calculated over its Voronoi region. The Voronoi diagram is updated after updating the positions and the probability map by all η agents.

Note that similar to the Q-max algorithm, cumulative rewards do not depend on the previous trajectories of the agents, and the process that governs the activity of each agent is a Markov process with states that include the positions of the agent and the probability maps. This property allows the use of the offline learning procedure in the collective Q-max algorithm as well.

In this process, at the step l, instead of the target location probabilities, the networks use the probabilities of the expected targets' locations $\Pr\{s(c_i, l) = 1 \mid s(c_i, l-1) = 1\}$ and $\Pr\{s(c_i, l) = 1 \mid s(c_i, l-1) = 0\}$ at step l given the states of the cells at the previous step $l-1$. For each kth search agent, these probabilities are defined by the same equations as in the Q-max algorithm of search by a single agent. Namely

$$\Pr\{s(c_i, l) = 1 \mid s(c_i, l-1) = 1\} = \frac{p_i(l-1)p_{\text{TA}}\exp[-d(c_i, c_j)/\lambda^k]}{p_i(l-1)(1-\alpha)+\alpha}$$
$$+ \frac{p_i(l-1)(1-\exp[-d(c_i, c_j)/\lambda^k])^2}{p_i(l-1)(1-p_{\text{TA}}\exp[-d(c_i, c_j)/\lambda^k]) + (1-p_i(l-1))(1-\alpha p_{\text{TA}}\exp[-d(c_i, c_j)/\lambda^k])}$$

and

$$\Pr\{s(c_i, l) = 1 \mid s(c_i, l-1) = 0\} = \frac{p_i(l-1)\alpha p_{\text{TA}}\exp[-d(c_i, c_j)/\lambda^k]}{p_i(l-1)(1-\alpha)+\alpha}$$
$$+ \frac{p_i(l-1)(1-\exp[-d(c_i, c_j)/\lambda^k])(1-\alpha p_{\text{TA}}\exp[-d(c_i, c_j)/\lambda^k])}{p_i(l-1)(1-p_{\text{TA}}\exp[-d(c_i, c_j)/\lambda^k]) + (1-p_i(l-1))(1-\alpha p_{\text{TA}}\exp[-d(c_i, c_j)/\lambda^k])}.$$

Then, at the learning stage, the targets' location probabilities are

$$p_i(l) = \begin{cases} \Pr\{s(c_i, l) = 1 \mid s(c_i, l-1) = 1\} & \text{if the target was in } c_i \text{ at } l-1, \\ \Pr\{s(c_i, l) = 1 \mid s(c_i, l-1) = 0\} & \text{if the target was not in } c_i \text{ at } l-1. \end{cases}$$

Termination of the algorithm is defined by the same as above conditions: learning process terminates when the updated probability map $P(l)$ becomes equal to the objective map P^*, and detection process terminates either when the updated probability map $P(t)$ becomes equal to the objective map P^* or when all ξ targets are detected.

The collective Q-max algorithm for detecting multiple targets also includes two stages: the learning stage when the agents' neural networks are trained, and the acting stage when the search agents detect the targets using the trained neural networks. The algorithm is outlined as follows.

Algorithm 7.2 Collective detection with deep learning: collective Q-max algorithm

---

Network structure:
input layer: $2n$ neurons (n agent positions and n target location probabilities with respect to the size n of the set C),
hidden layer: $2n$ neurons,
output layer: 9 neurons (with respect to the number of actions – size of the set $\mathbb{A}$).

Activation function:
sigmoid function $f(x) = 1/(1+e^{-x})$.

Loss function:
mean square error (MSE) function.

Input: Set $C = \{c_1, c_2, \dots, c_n\}$,
number of agents η,
sensor sensitivities $\lambda^1, \lambda^2, \dots, \lambda^\eta$
initial agents' positions $c^1(0), c^2(0), \dots, c^\eta(0)$,
set $\mathbb{A} = \{\uparrow, \nearrow, \rightarrow, \searrow, \downarrow, \swarrow, \leftarrow, \nwarrow, \odot\}$ of possible movements,
probability p_{TA} of true alarms,
rate α of false alarms and probability $p_{\text{FA}} = \alpha p_{\text{TA}}$,
initial probability map $P(0) = \{p_1(0), p_2(0), \dots, p_n(0)\}$ on the domain C,
value ε,
objective probability map P^* (obtained by using the value ε),
number of targets ξ,

Output: target locations $\hat{c}^1(T), \hat{c}^2(T), \dots, \hat{c}^{\xi}(T)$ at a termination time T.

Learning

1. Generate training data set (c, P) with
    agents' positions $c(0) = (c^1(0), c^2(0), \dots, c^{\eta}(0))$ and
    probability map $P(0) = \{p_1(0), p_2(0), \dots, p_n(0)\}$.
2. For each agent $k = 1, \dots, \eta$ do:
3. Create the prediction network.
4. Create the target network as a copy of the prediction network.
5. End for.
6. Start with $l = 0$.
7. For each pair (c, P) from the training dataset, do:
8. Create the Voronoi diagram $\mathfrak{C}(l) = \{C^1(l), C^2(l), \dots, C^{\eta}(l)\}$.
9. Create the probability atlas $\mathfrak{P}(l) = \{P^1(l), P^2(l), \dots, P^{\eta}(l)\}$.
10. For each agent $k = 1, \dots, \eta$ do:
11. For each action $\mathbb{a}^k \in \mathbb{A}$ do:
12. Calculate the value $Q(c^k(l), P(l), \mathbb{a}^k(l))$ by the prediction network.
13. Choose action $\mathbb{a}^k(l)$ by the value $p(\mathbb{a}|Q;\theta)$ and the SoftMax policy.
14. Apply the chosen action to the current position $c^k(l)$ and obtain the next position $c^k(l+1)$.
15. Update the probability map $P(l)$ to $P(l+1)$.
16. If $P(l+1) = P^*$, then
17. Set immediate reward $R^k_{\mathbb{a}}(l) = 0$.
18. Set cumulative reward $Q(c^k(l), P(l), \mathbb{a}^k(l), w) = 0$.
19. Else
20. Calculate immediate reward $R^k_{\mathbb{a}}(l)$ with respect to probabilities the agent's parts $P^k(l) \in \mathfrak{P}(l)$ and $P^k(l+1) \in \mathfrak{P}(l)$ of the probability map. The Voronoi diagram and so – the cells associated with the kth agent remain, but the values of the probabilities change.
21. End if.
22. End for.
23. Calculate the values $Q^+(c^k(l+1), P(l+1), \mathbb{a}^k; w')$ by the target network.
24. Calculate the temporal difference learning error $\Delta_l(Q)$ for maximal Q^+.
25. Update the prediction network with respect to the error $\Delta_l(Q)$.
26. Update the target network by the weights of the prediction network.
27. Update the prediction network of the $(k+1)$th agent by the target network of the kth agent.
28. End for.
29. If the training epochs ended
30. For each agent $k = 1, \dots, \eta$ do:
31. Update the target network by the target network of the ηth agent.
32. End for.
33. Start acting (go to line 36).
34. End if.
35. End for.

Acting

36. Start with $t = 0$.

37. Obtain the initial agents' positions $c^1(t), c^2(t), \ldots, c^\eta(t)$.
38. Obtain the initial probability map $P(t) = \{p_1(t), p_2(t), \ldots, p_n(t)\}$.
39. For each agent $k = 1, \ldots, \eta$ do:
40. Get the values $Q(c^k(t), P(t), \mathbb{a}^k(t), w)$ using the trained network.
41. Choose action $\mathbb{a}^k(t)$, which provides the maximum $Q(c^k(t), P(t), \mathbb{a}^k(t), w)$.
42. Apply the chosen action to the current position $c^k(t)$ and obtain the next position $c^k(t+1)$.
43. Screen the domain C. {The kth agent screens all the domain C with respect to the abilities of the on-board sensors.}
44. Update the targets' locations $\widehat{c}^1(t), \widehat{c}^2(t), \ldots, \widehat{c}^\xi(t)$.
45. Update the probability map $P(t)$ to $P(t+1)$.
46. End for.
47. If all ξ targets were detected (or if $P(t) = P^*$), then
48. Set $T = t$ and terminate (go to line 53).
49. Else
50. Set $t = t + 1$.
51. Continue detection (go to line 37).
52. End if.
53. Return: targets' locations $\widehat{c}^1(T), \widehat{c}^2(T), \ldots, \widehat{c}^\xi(T)$.

The presented algorithm extends a previously developed detection algorithm by a single agent and follows the same techniques for training the networks. However, instead of using individual prediction and target networks, it considers the prediction network of the next agent as its target network. Thus, each agent uses the training results of the previous agent and shares its knowledge with the next agent.

7.3 Numerical Simulations

Numerical simulations were implemented using Python programming language, including the PyTorch machine learning library. The trials were executed on a PC Intel® Core™ i7-10700 CPU with 16 GB RAM (8 cores) with a GPU Nvidia GeForce GTX 1650 Super (1280 CUDA Cores). Using this computer system, we measured the run time of the simulations over different datasets, demonstrating that the suggested algorithms can be implementable on conventional computers with CUDA parallel computing architecture and do not require specific parameters for their functionality.

In the collective Q-max algorithm, the initial weights w of neural networks were generated by the corresponding procedures of the PyTorch library. The optimizer used in the simulation was the ADAM optimizer from the PyTorch library. The training data set size was 100,000, the number of training epochs was 30, and the average time required for offline training of the prediction network was approximately 10 hours on the described computation platform. After an offline training period, online decision-making was conducted directly by applying immediate selection without additional calculations.

In the simulations, we compared four algorithms: random search, in which the agents move randomly in the domain, centralized and distributed EIG algorithms and collective Q-max algorithm.

In all algorithms, we used the sizes 40×40 and 50×50 cells grid, which corresponds to practical military and civil tasks. In the case of military applications requiring search and detection in active operations, the size of the cell in the detection tasks is up to 0.5 km^2; thus, the grid of the indicated size represents a city district or a natural terrain of up to 1.0 km^2. In the case of civil applications, for example, in the search and rescue tasks and military logistic tasks, such grids represent a terrain of more than 6.0 km^2 while searching the objects of the size of a human and up to 25.0 km^2 while searching the objects of the size of an automobile.

Sensitivity of the sensor is specified by the parameter λ and in the simulations we used the values $\lambda = 10$ and $\lambda = 15$ that are associated with the sensitivity of the lidars, where the probability of detecting the target decreases from 1 (detection at the distance of 0 m) to $1/e$ (detection at the distance of 15 m). Certainly, in the real-world tasks the sensors' sensitivity λ should be defined with respect to the type and quality of the sensor.

7.3.1 Detection of Static Targets

The first set of simulations dealt with the detection of static targets. The results of these simulations are illustrated by the detection of $\xi = 30$ targets by $\eta = 6$ agents in the domain of the size $n = 40 \times 40 = 1600$ cells. The probability of a true alarm is $p_{TA} = 1$, the sensor sensitivity is $\lambda = 15$, the initial target location probabilities are $p_i(0) = 0.05$ and $i = 1, 2, \dots, n$, and the initial agent positions are $c^k(0) = (0, 0)$ and $k = 1, 2, \dots, \eta$.

Dependence of the cumulative reward on time for the ratio of false alarms $\alpha = 0.5$ is shown in Figure 7.5.

The collective Q-max Algorithm 7.2 provides the fastest growth of the cumulative reward, and the slowest growth is provided by the random search method. The random search is slightly outperformed by the grid search, in which the paths of the agents are determined in advance; at the beginning of the search process at $t = 0$, the domain is divided into equal areas with respect to the number of agents, and each agent conducts the center of gravity search in its area according to the initial probability map. Finally, intermediate growth is provided by the EIG algorithms, where the distributed EIG Algorithm 7.1 outperforms the centralized EIG algorithm (see Chapter 4).

The same tendency was observed for the number of search actions conducted by the algorithms up to detecting all the targets. The results of these simulations with different α ratios of false alarms, where the sensitivity of the agents' sensors is defined by $\lambda = 15$, are summarized in Table 7.1.

As was expected, the worst results were obtained for a random search, in which the agents move randomly in the domain, while the collective Q-max algorithm provides the best results. Note that

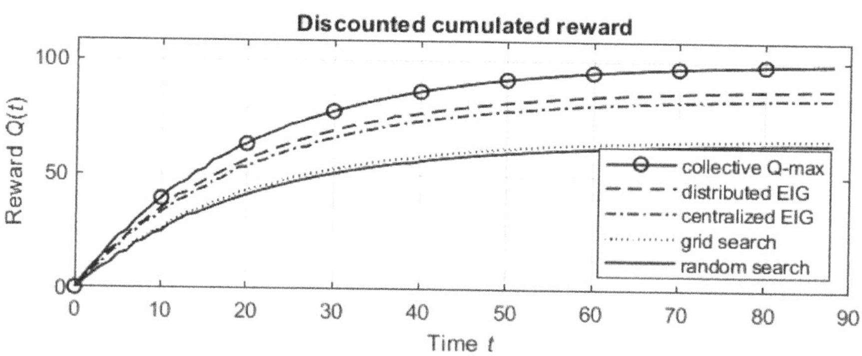

Figure 7.5 Dependence of the discounted cumulative reward on time; the ratio of the false alarms is $\alpha = 0.5$.

Table 7.1 The number of actions up to the detection of 30 static targets by 6 agents for different ratios of false alarms.

	Number of actions up to detection of 30 static targets $n = 40 \times 40$, $\lambda = 15$		
Detection algorithm	$\alpha = 0.25$	$\alpha = 0.5$	$\alpha = 0.75$
Random search	120	175	450
Grid search	112	170	436
Centralized EIG	88	122	232
Distributed EIG	79	98	152
Collective Q-max	**75**	**88**	**102**

The sensors' sensitivity is $\lambda = 15$ and the domain size is 40×40. The minimal numbers of actions are stressed by bold.

the difference between the best results provided by the collective Q-max algorithm and the other methods increases with the ratio of false alarms.

For comparison, in Table 7.2 we present the number of search actions executed by the algorithms to detect all the targets with a lower sensitivity $\lambda = 10$.

Since sensitivity $\lambda = 10$ of the agents' sensors is lower than that in the previous scenario, the agents need more search actions to detect the targets.

To stress the difference between the considered search algorithms, Figure 7.6 illustrates the effectiveness of the algorithms in comparison with collective Q-max, which's efficiency is defined as 100%.

The figure shows that the collective Q-max algorithm outperforms the other algorithms, and its effectiveness relative to the other algorithms is as high as the sensors' sensitivity is lower. For example, for $\lambda = 15$ and $\alpha = 0.25$, the collective Q-max algorithm requires nearly 2 times less time for detecting the targets than the random search, and for $\lambda = 15$ and $\alpha = 0.75$, it requires nearly 3 times less time than the random search. Similar ratios are observed for the other algorithms and values of λ and α.

Table 7.2 The number of search actions up to the detection of 30 static targets by 6 agents for different ratios of false alarms.

	Number of actions up to detection of 30 static targets $n = 40 \times 40$, $\lambda = 10$		
Detection algorithm	$\alpha = 0.25$	$\alpha = 0.5$	$\alpha = 0.75$
Random search	205	310	>600
Grid search	198	305	590
Centralized EIG	145	221	321
Distributed EIG	134	210	285
Collective Q-max	**106**	**138**	**178**

Sensor sensitivity is $\lambda = 10$; the domain size is 40×40. The minimal numbers of actions are stressed by bold.

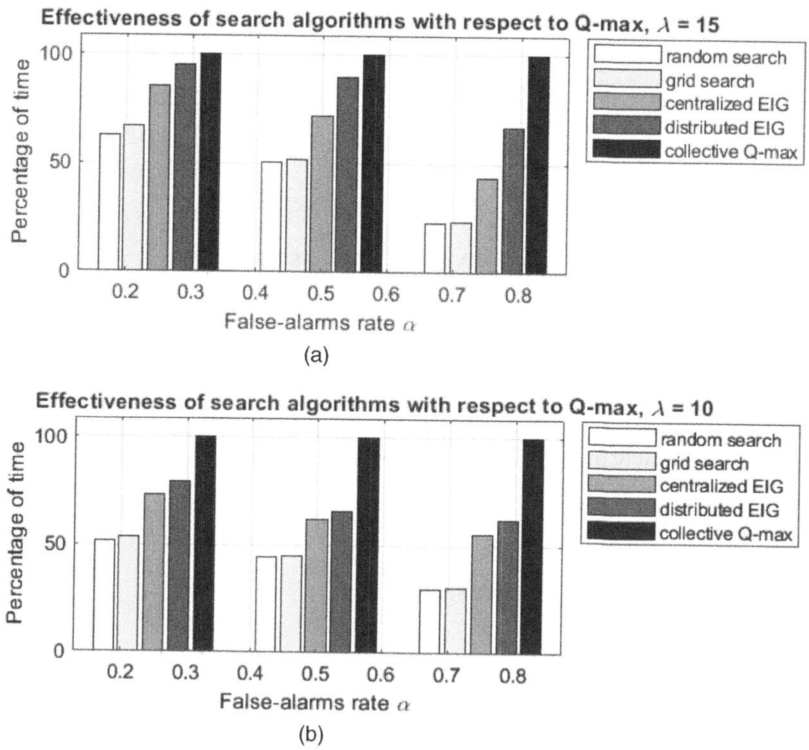

Figure 7.6 Effectiveness of search algorithms in the search of 30 static targets by 6 agents with two values of sensors' sensitivity λ: (a) $\lambda = 15$ and (b) $\lambda = 10$. The size of the domain is $n = 40 \times 40$.

Table 7.3 The number of search actions up to the detection of 100 static targets by 10 agents for different ratios of false alarms.

	Number of actions up to detection of 100 static targets $n = 50 \times 50$, $\lambda = 10$		
Detection algorithm	$\alpha = 0.25$	$\alpha = 0.5$	$\alpha = 0.75$
Random search	202	275	>600
Grid search	196	272	>600
Centralized EIG	156	215	322
Distributed EIG	135	198	305
Collective Q-max	**102**	**135**	**165**

Sensor sensitivity is $\lambda = 10$; the domain size is 50×50. The minimal numbers of actions are stressed by bold.

The same tendencies were demonstrated in other scenarios. For example, the number of actions conducted by the algorithms up to detecting 100 static targets by 10 agents equipped with sensors of sensitivity $\lambda = 10$ are presented in Table 7.3; the domain size is $n = 50 \times 50$.

Similar to the above, in this scenario, the collective Q-max algorithm leads to essentially better results than the other algorithms, and its priority is shown to be as high as the ratio of false alarms. Therefore, when considering the detection of static targets, the suggested collective Q-max algorithm is preferable to the other methods, and such preference is more significant in tasks with a high ratio of false-positive errors. On the other hand, if the ratio of false-positive errors is relatively low, then the suggested distributed EIG algorithm can also be applied.

7.3.2 Detection of Moving Targets

The second set of simulations dealt with the detection of moving targets. As above, we present the results of simulations with $\xi = 30$ targets searched by $\eta = 6$ agents in a domain of size $n = 40 \times 40 = 1600$ cells.

The probability of a true alarm is $p_{\text{TA}} = 1$, the sensor sensitivity is given by $\lambda = 15$ and the initial targets' location probabilities are uniformly distributed with $p_i(0) = 0.05$, $i = 1, 2, \ldots, n$, while the initial agents' positions are $c^k(0) = (0, 0)$, $k = 1, 2, \ldots, \eta$.

For simplicity, here, we consider slowly moving targets such that the probability of staying in the same cell is substantial,

$$Pr\{\hat{\mathbb{a}}(t) = \odot\} = 0.9$$

and the probability of taking a search step in any direction is

$$Pr\{\hat{\mathbb{a}}(t) \in \mathbb{A}\backslash\odot\} = \frac{1 - Pr\{\hat{\mathbb{a}}(t) = \odot\}}{\#\mathbb{A} - 1} = 0.0125.$$

For convenience, we consider the results of the simulations with the same sensor sensitivities $\lambda = 15$ and $\lambda = 10$ and three ratios of false alarms.

The false-alarm ratios are $\alpha = 0.1$, $\alpha = 0.15$ and $\alpha = 0.25$ because in scenarios with higher false-alarm ratios detection process is longer and the resulting high numbers of search actions, despite the same relations between them, are less illustrative.

The results of the simulations with different false-alarm ratios α and a fixed sensitivity of the agents' sensors are given by $\lambda = 15$. These results are summarized in Table 7.4.

It is seen that the worst results are obtained by random search and that the collective Q-max algorithm provides the best results. Note that for moving targets, when $\alpha = 0.25$, the random search method requires 6 times more actions (720) than in the case of detecting static targets (120, see Table 7.1).

Table 7.4 The number of search actions up to detecting 30 moving targets by 6 agents for different ratios of false alarms.

	Number of actions up to detection of 30 moving targets $n = 40 \times 40$, $\lambda = 15$		
Detection algorithm	$\alpha = 0.1$	$\alpha = 0.15$	$\alpha = 0.25$
Random search	205	310	720
Grid search	190	301	710
Centralized EIG	140	208	366
Distributed EIG	132	180	340
Collective Q-max	**115**	**132**	***188***

Sensor sensitivity is given by $\lambda = 15$. The size of the domain is 40×40. The minimal numbers of actions are stressed by bold.

Table 7.5 The number of search actions required for detecting 30 moving targets by 6 agents for different probabilities of false alarms.

	Number of actions up to detection of 30 moving targets $n = 40 \times 40$, $\lambda = 10$		
Detection algorithm	$\alpha = 0.1$	$\alpha = 0.15$	$\alpha = 0.25$
Random search	280	410	850
Grid search	268	405	844
Centralized EIG	202	245	475
Distributed EIG	185	225	450
Collective Q-max	**128**	**155**	**220**

The sensors' sensitivity is given by $\lambda = 10$; the size of the domain is 40×40. The minimal numbers of actions are stressed by bold.

Table 7.6 The number of search actions required up to the detection of 100 moving targets by 10 agents for different ratios of false alarms.

	Number of actions up to detection of 100 moving targets $n = 50 \times 50$, $\lambda = 10$		
Detection algorithm	$\alpha = 0.1$	$\alpha = 0.15$	$\alpha = 0.25$
Random search	295	422	865
Grid search	285	411	861
Centralized *EIG*	205	247	465
Distributed *EIG*	190	232	447
Collective Q-max	**132**	**162**	**225**

The sensor sensitivity is $\lambda = 10$, and the domain size is 50×50. The minimal numbers of actions are stressed by bold.

For the same α value, the EIG algorithm requires nearly 4 times more search actions (366 vs. 88 and 340 vs. 79). Finally, the ratio between these numbers of required actions between the static and the dynamic case when applying the suggested collective Q-max algorithm is only $\frac{188}{75} \approx 2.5$.

For comparison, in Table 7.5 we present the number of search actions conducted by the algorithms up to detection of all the targets with the lower sensitivity $\lambda = 10$.

As was expected, since the sensitivity of the agents' sensors is given by $\lambda = 10$, the ratios between the number of required actions are lower compared to this ratio in the previous scenario, and the agents need more actions to detect the targets. The number of search actions follows the same tendency as above and the collective Q-max algorithm strongly outperforms the other methods.

Figure 7.7 illustrates the effectiveness of the algorithms in comparison with the collective Q-max algorithm, which's efficiency is defined as 100% (cf. Figure 7.6).

Finally, parallel to the results presented in Table 7.3, the number of actions executed by the algorithms up to the detection of 100 moving targets by 10 agents equipped with sensors of sensitivity $\lambda = 10$ are presented in Table 7.6. The size of the search domain is $n = 50 \times 50$.

In this scenario, the collective Q-max algorithm obtains the best results in comparison with the other algorithms, and its advantage is higher as the ratio of false alarms increases.

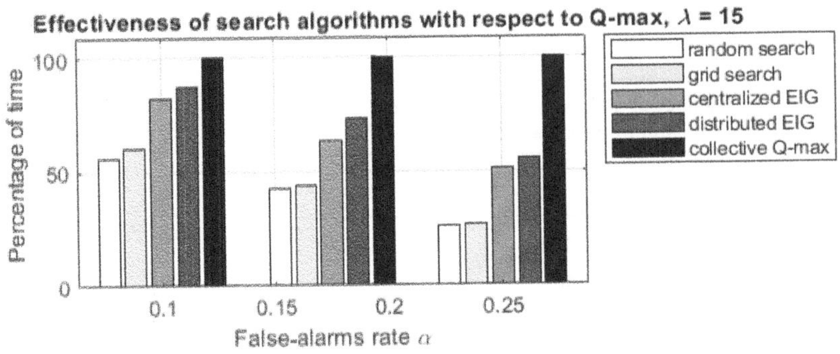

(a)

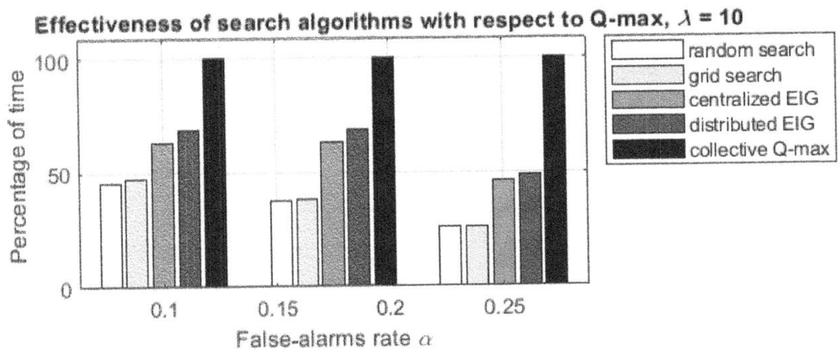

(b)

Figure 7.7 Effectiveness of search algorithms in the search of 30 moving targets by 6 agents with two values of sensors' sensitivity λ: (a) $\lambda = 15$ and (b) $\lambda = 10$. The domain size is $n = 40 \times 40$.

7.3.3 Learning Errors and Run Time of the Collective Q-max Algorithm

We consider two characteristics of the collective Q-max algorithm which emphasize the feasibility of its implementation in solving real-world problems.

We used 30 learning epochs with 100,000 training data sets in the presented simulations. This choice of parameters is based on the dependence of the percentage of learning errors on the number of learning epochs. The graph of this dependence is shown in Figure 7.8.

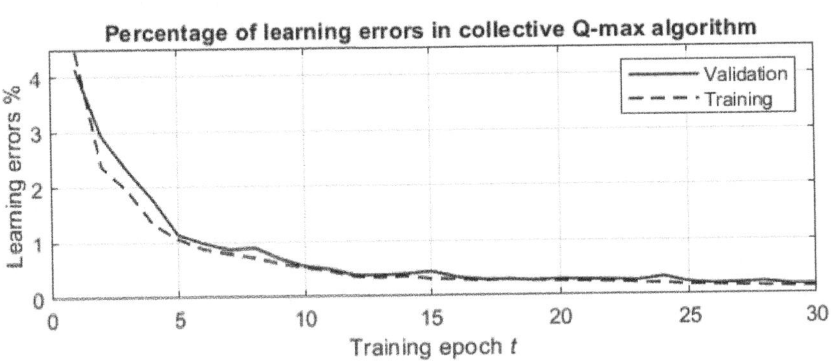

Figure 7.8 Dependence of the percentage of learning errors on the number of training epochs.

Table 7.7 The times required for training the neural networks for different domains and datasets sizes.

Domain size $n_x \times n_y$	Number of nonzero weights in the neural network	Size of the data set	Run time for one epoch (minutes)	Mean squared error[a)]
20×20	648,009	50,000	5	0.20
		100,000	9	0.13
40×40	10,272,009	50,000	12	0.28
		100,000	20	0.17
50×50	25,050,009	50,000	14	0.30
		100,000	22	0.19

a) The error was calculated over the temporal difference errors at the validation stage at epoch $t = 30$.

The percentage of learning errors decreases exponentially and converges to 0.1% after 30 training epochs.

The run times required for training the neural networks in different simulation settings for the abovementioned computation system are summarized in Table 7.7.

It is seen that while the size of the domain and the number of links in the network increase exponentially, the mean squared error increases very slowly and remains relatively small.

Note that even on a simple PC system described above, computations require reasonable time. After training decision-making is processed using the already trained networks that allow the application of the considered techniques in online algorithms.

7.4 Conclusions

In this chapter, we considered two algorithms of search for static and moving targets by a group of mobile search agents.

The first algorithm, DEIG, is a reactive online procedure in which each agent observes the environment and makes its decision using the obtained EIG value. This algorithm extends the centralized EIG algorithm considered in Chapter 4.

The DEIG algorithm uses Voronoi diagrams in parallel with the probability map. As a result, the search efforts are distributed in such a manner that the agents first consider their neighborhoods and then continue with the detection in the other regions.

The main advantage of the DEIG algorithm is its computational simplicity and short run time. The conducted simulation studies demonstrate the algorithm's effectiveness, especially in cases with a low ratio of detection errors of the second type.

The second algorithm, the collective Q-max algorithm, includes deep Q-learning abilities that can be used online and offline. The algorithm extends the Q-max algorithm (see Chapter 6) to a team of the agents.

The basic idea of the used learning procedure is the sequential training of the prediction neural network. Each search agent is trained by its predecessor in the array of the agents and trains its successor by simultaneously counting the expected actions.

Training can be conducted both offline and online or can be processed in a mixed regime with online updating of the offline training results. In the considered scenarios, the order of the agents varies and is specified with respect to the distance of the agent from the starting point. Such definition allows to include in the search all agents in minimal time.

In the other scenarios, the order can be predefined by enumeration of the agents or can vary with respect to information measures or distances between the agents.

The collective Q-max algorithm considerably outperforms the benchmark algorithms, including the distributed EIG algorithm. The main advantages of the algorithm are demonstrated in a noisy environment and in the use of sensors with low sensitivity with many statistical errors of the first and second types.

In addition, simulations show that both Q-max algorithms outperform greedy and learning procedures and require reasonable and relatively moderate computation time that makes them useful in different applications.

8

Conclusions

We considered the online reactive and learning algorithms of search for multiple static and moving targets by a group of mobile agents. The algorithms effectively solve the problem of search and govern the search agents in the presence of false-positive and false-negative detection errors.

The algorithms follow classical Koopman formulation applied to the search by a finite number of search agents acting in a gridded two-dimensional domain. The knowledge about the locations of the targets is represented by the probability map updated by the agents during the search.

Reactive algorithms govern the online activity of a single agent or a group of communicating agents. These algorithms utilize the expected information gain (EIG) and implement the center of gravity (COG) and center of view (COV) strategies for planning one-step or several steps motions.

These strategies are implemented for both the complete domain and the Voronoi regions created by dynamical Voronoi partitioning of the search domain. The best results in the sense of search time were reached by the algorithm that maximizes one-step EIG in each Voronoi region.

The algorithms with deep Q-learning use the prediction and target neural networks for learning the targets' activity. In the search by a single agent, both networks are trained, updated, and used conventionally by the same agent. In the collaborative search, we use the round learning scheme in which the agents communicate such that the next-in-the-line agent updates its prediction network with respect to the target network of the previous agent. The networks can be trained offline with further online correction using the sensed information or online starting with certain initial probability maps.

The collective Q-learning algorithm uses the learned information about the targets' location probabilities, the detection errors, and the targets' motion. In the simulations, the targets' motion was characterized as 90% static (remaining stationary) and 10% as a random walk without any specific movement patterns. If the targets move according to certain movement patterns, such as attempting to evade the search agents, the collective Q-learning algorithm is able to recognize and learn these patterns, resulting in a shorter search time.

Simulations demonstrated that the Q-learning algorithms effectively solve the problem of search and detection by the group of mobile agents in the presence of false-positive and false-negative detection errors and considerably outperform the known methods.

References

1 Alpern, S. and Gal, S. (2003). *The Theory of Search Games and Rendezvous*. New York: Kluwer Academic Publishers.

2 Albrecht, S., Christianos, F., and Schäfer, L. (2024). *Multi-Agent Reinforcement Learning. Foundations and Modern Approaches*. Cambridge, MA: The MIT Press.

3 Bertsekas D. *Multiagent Value Iteration Algorithms in Dynamic Programming and Reinforcement Learning*. 4 May 2020, arXiv:2005.01627v1.

4 Brown, S. (1980). Optimal search for a moving target in discrete time and space. *Operations Research* 28 (6): 1275–1289.

5 Chernikhovsky G., Kagan E., Goren G., Ben-Gal I. Path planning for sea vessel search using wideband sonar. In *Proc. 27th IEEE Convention of Electrical and Electronics Engineers in Israel*, 14–17 Nov 2012, Eilat.

6 Ciervo, A. (1976). *Search for Moving Targets*. PSR Report 619. Santa-Monica, CA: Pacific-Sierra Research Corporation.

7 Cooper, D., Frost, J., and Quincy, R. (2003). *Compatibility of Land SAR Procedures with Search Theory*. Washington, DC: US Department of Homeland Security.

8 Cover, T. and Thomas, J. (1991). *Elements of Information Theory*. New York: John Wiley & Sons.

9 Dai W., Sartoretti G. Multi-agent search based on distributed deep reinforcement learning. In *Proc. The 3rd Asian Conf. Artificial Intelligence Technology ACAIT 2019*, 5–7 Jul 2019, Chongqing, China.

10 Dames, P. (2020). Distributed multi-target search and tracking using the PHD filter. *Autonomous Robots* 44: 673–689.

11 Eagle, J. and Washburn, A. (1991). Cumulative search-Evasion games. *Naval Research Logistics* 38: 495–510.

12 Elfes, A. (1987). Sonar-based real-world mapping, and navigation. *IEEE J. Robotics and Automation* 3 (3): 249–265.

13 Elfes A. Occupancy grids: a stochastic spatial representation for active robot perception. In *Proc. 6th Conference on Uncertainty in Artificial Intelligence*, 1990, 136–146.

14 Filar, J. and Vrieze, K. (1997). *Competitive Markov Decision Processes*. New York: Springer-Verlag.

15 Frost, J. and Stone, L. (2001). *Review of Search Theory: Advances and Applications to Search and Rescue Decision Support*. Groton, CT: US Coast Guard Research and Development Center.

16 Gal, S. (1988). *Search Games*. New York: Academic Press.

17 Garaffa, L., Basso, M., Konzen, A., and de Freitas, E. (2023). Reinforcement learning for mobile robotics exploration: a survey. *IEEE Trans. Neural Networks and Learning Systems* 34 (8): 3796–3810.

18 Garnaev, A. (2000). *Search Games and Other Applications of Game Theory*. Berlin, Germany: Springer-Verlag.

19 GeekForGeeks. *Actor-Critic Algorithm in Reinforcement Learning*. https://www.geeksforgeeks.org/machine-learning/actor-critic-algorithm-in-reinforcement-learning. 2025. Last accessed 11.03.2026.

20 GitHub. *Cooperative Search by Mobile Agents under Uncertainty*. https://github.com/Barouchm/SearchProblem. 2023. Last accessed 11.03.2026.

21 Hacohen, S., Shoval, S., and Shvalb, N. (2019). Probability navigation function for stochastic static environments. *Int. J. Control, Automation and Systems* 17 (8): 2097–2113.

22 Hacohen S., Shoval S., Shvalb N. Multi agents multi targets mission under uncertainty using probability navigation function. In *Proc. 13th IEEE International Conference on Control & Automation ICCA'17*, 2017, 845–850.

23 Hacohen, S., Shoval, S., and Shvalb, N. (2017). Applying probability navigation function in dynamic uncertain environments. *Robotics and Autonomous Systems* 87: 237–246.

24 Hespanha J., Prandini M., Sastry S. Multi-agent probabilistic pursuit-evasion games. In *Proc. 38-th IEEE Conference on Decision Making and Control*, 3, 1999, 2432–2437.

25 Hespanha J., Prandini M., Sastry S. Probabilistic pursuit-evasion games: a one-step Nash approach. In *Proc. 39-th IEEE Conference on Decision Making and Control*, 2000, 2272–2277.

26 Igoe, C., Ghods, R., and Schneider, J. (2022). Multi-agent active search: a reinforcement learning approach. *IEEE Robotics and Automation Letters* 7 (2): 754–761.

27 Isaacs, R. and Karlin, S. (1954). *A Game of Aiming and Evasion*. Research Memorandum, 1316. Santa Monica, CA: RAND Corporation.

28 Isaacs, R. (1965). *Differential Games: A Mathematical Theory with Applications to Warfare and Pursuit, Control and Optimization*. John Wiley & Sons: New York, London, Sydney.

29 Israel M., Khmelnitsky E., Kagan E. Search for a mobile target by ground vehicle on a topographic terrain. In *Proc. 27th IEEE Convention of Electrical and Electronic Engineers in Israel*, 14–17 Nov 2012, Eilat.

30 Jeong H., Hassani H., Morari M., Lee D., Pappas G. *Learning to Track Dynamic Targets in Partially Known Environments*, 17 Jun 2020, arxiv.org/abs/2006.10190v1.

31 Kaelbling, L., Littman, M., and Moore, A. (1996). Reinforcement learning: a survey. *J. Artificial Intelligence Research* 4: 237–285.

32 Kagan E., Ben-Gal I. An informational search for a moving target. In *Proc. IEEE 24th Convention of Electrical and Electronics Engineers in Israel*, 15–17 Nov. 2006, Eilat.

33 Kagan E., Goren G., Ben-Gal I. Probabilistic double-distance algorithm of search after static or moving target by autonomous mobile agent. In *Proc. IEEE 26th Convention of Electrical and Electronics Engineers in Israel*, 17–20 Nov, 2010, Eilat.

34 Kagan E., Goren G., Ben-Gal I. Algorithm of search for static or moving target by autonomous mobile agent with erroneous sensor. In *Proc. IEEE 27th Convention of Electrical and Electronics Engineers in Israel*, 14–17 Nov, 2012, Eilat.

35 Kagan, E. and Ben-Gal, I. (2013). *Probabilistic Search for Tracking Targets. Theory and Modern Applications*. Chichester, UK: John Wiley & Sons.

36 Kagan, E. and Ben-Gal, I. (2015). *Search and Foraging. Individual Motion and Swarm Dynamics*. Boca Raton, FL: Taylor & Francis.

37 Kagan, E., Shvalb, N., and Ben-Gal, I. (ed.) (2019). *Autonomous Mobile Robots and Multi-Robot Systems: Motion-Planning, Communication and Swarming*. Chichester, UK: Wiley & Sons.

38 Kober, J. and Peters, J. (2014). *Learning Motor SkillsFrom Algorithms to Robot Experiments.* Cham, Switzerland: Springer.

39 Koopman, B. (1946). *Search and Screening*. Operation Evaluation Research Group Report, 56. Arlington, VA: Center for Naval Analysis.

40 Koopman, B. (1956). The theory of search. *Operations Research* 4, 324–346, 503–531, 5, 1957, 613–626.

41 Koopman, B. (1980). *Search and Screening: General Principles with Historical Applications.* New York: Pergamon Press.

42 Laperriere-Robillard, T., Morin, M., and Abi-Zeid, I. (2022). Supervised learning for maritime search operations: an artificial intelligence approach to search efficiency evaluation. *Expert Systems with Applications* 206: 117857.

43 Nahin, P. (2007). *Chases and Escapes: The Mathematics of Pursuit and Evasion*. Princeton, NJ: Princeton University Press.

44 Nguyen T., Nguyen N., Nahavand S. *Deep Reinforcement Learning for Multi-agent Systems: A Review of Challenges, Solutions, and Applications.* 6 Feb 2019, arxiv.org/abs/1812.11794v2.

45 Ohsumi, A. (1991). Optimal search for a Markovian target. *Naval Research Logistics* 38: 531–554.

46 Piacentini, C., Bernardini, S., and Beck, J. (2019). Autonomous target search with multiple coordinated UAVs. *J. Artificial Intelligence Research* 65: 519–568.

47 Pollock, S. (1970). A simple model of search for a moving target. *Operations Research* 18: 883–903.

48 Prasuna, R. and Potturu, S. (2024). Deep reinforcement learning in mobile robotics – a concise review. *Multimedia Tools and Applications* 83: 70815–70836.

49 Quiroga, F., Hermosilla, G., Farias, G. et al. (2022). Position control of a mobile robot through deep reinforcement learning. *Applied Science* 12: 7194.

50 Robin, C. and Lacroix, S. (2016). Multi-robot target detection and tracking: taxonomy and survey. *Autonomous Robots* 40 (4): 729–760.

51 Ross, S. (1969). A problem in optimal search and stop. *Operations Research* 17 (6): 984–992.

52 Ross, S. (1983). *Introduction to Stochastic Dynamic Programming*. New York: Academic Press.

53 Sutton, R. and Barto, A. (1998). *Reinforcement Learning: An Introduction*, 2nde. Cambridge, MA: Bradford Book, MIT Press.

54 Senanayake, M., Senthooran, I., Barca, J. et al. (2016). Search and tracking algorithms from swarm of robots: a survey. *Robotics and Autonomous Systems* 75: 422–434.

55 Stone, L. (1975). *Theory of Optimal Search*. New York: Academic Press.

56 Stone, L., Barlow, C., and Corwin, T. (1999). *Bayesian Multiple Target Tracking*. Boston, MA: Artech House Inc.

57 Stone, L., Royset, J., and Washburn, A. (2016). *Optimal Search for Moving Targets*. Cham, Switzerland: Springer.

58 Sung Y., Budhiraja A., Williams R., Tokekar P. Distributed simultaneous action and target assignment for multi-robot multi-target tracking. In *Proc. IEEE International Conference on Robotics and Automation ICRA' 2018*, 21–25 May 2018, Brisbane, Australia, 3724–3729.

59 Washburn, A. (1980). On a search for a moving target. *Naval Research Logistics Quarterly* 27: 315–322.

60 Washburn, A. (1983). Search for a moving target: the FAB algorithm. *Operations Research* 31 (4): 739–751.

61 Washburn, A. (1989). *Search and Detection*. Arlington, VA: ORSA Books.

62 Washburn, A. (1995). *Branch and Bound Methods for Search Problems.* Technical Report 1995-04,. Monterey, CA: Naval Postgraduate School.

63 Washburn, A. and Kress, M. (2009). *Combat Modelling*. Dordrecht, the Netherlands: Springer.

64 White, D. (1993). *Markov Decision Processes.* Chichester, UK: John Wiley & Sons.

65 Yadin, M. and Zacks, S. (1988). Visibility probabilities on line segments in three-dimensional spaces subject to random Poisson fields of obscuring spheres. *Naval Research Logistics* 35: 555–569.

66 Zacks, S. and Yadin, M. (1994). The survival probability function of a target moving along a straight line in a random field of obscuring elements. *Naval Research Logistics* 41: 689–706.

67 Zhu, C., Dastani, M., and Wang, C. (2024). A survey of multi-agent deep reinforcement learning with communication. *Autonomous Agents and Multi-Agent Systems* 38: 4.

Index

a

Accuracy 12, 48, 50
Action 50
  selection (learning stage) 58, 63
Actor 52–54
  actor-critic 52–54
Additive information property 32
Agent probability map 21–25, 31–33
Alarm
  false 22, 26, 27, 29–30
  true 22, 26, 27, 29–30
Algorithm
  collective Q-max 81
  DEIG (distributed EIG) 77, 80, 81, 88–93, 96
  EIG (expected information gain) 3, 32, 44, 55, 69, 70, 76
  Q-max 55, 58, 59, 61–73, 81
  search in shadowed space 10–16, 19
  search with erroneous detection 6, 16, 17, 19, 21, 22, 26, 30
  search with errorless detection 7–10
  SPL (shortest path length) 55, 58, 67–75
  visibility analysis 12–13, 16
Antisubmarine Warfare Operations Research Group 1

b

Bayesian sensor fusion scheme 6, 21, 23, 49
Bellman equation 44, 60, 67, 83
Bio-mimetic search 4
Brute-force learning 34–38, 43
Brute force search 16

c

Cell 6
  dominated 12
  intermediate 13
  potential 11–15
Center of Gravity (COG) 10, 18, 32, 34, 36–42, 44, 70, 71, 72
Center of View (COV) 32–42, 44, 70–73
Central station (unit) 5, 21, 24, 27
Collective Deep Q-Learning 81
Collision avoidance 5
Communication between agents 5, 26–27
Continuous space and time 4
Cooperative detection (Voronoi regions) 77–82
Cooperative search 5, 30, 78
Critic 50–53

d

Decision-making 5, 6, 8, 9, 17, 31, 35, 38, 46, 58, 61, 68, 79, 80, 88
  policies 39, 58
Decision process, Markov (MDP) 44, 52
Deep reinforcement learning 47
Detection 2, 3
  erroneous 16, 17, 43
  error functions 16
  errorless 7, 16, 19
  errors (false-negative) 16, 39, 55, 56
  errors (false-positive) 16, 18, 22, 25, 30, 39, 55, 56, 73, 92
  problem 5, 44, 47, 52, 58, 64, 77, 81
  uncertain 17
Diffusion processes 11, 17
Distributed EIG algorithm 79–81, 88–93

Domain 2
gridded 6, 21, 28, 31, 51, 52, 78
search 3, 48, 50, 77, 93
shadowed (shadowed space) 6, 10, 14, 27
Dominance (visibility analysis) 12
Dummy targets 22
Dynamic programming 55, 59, 60, 73, 74
Dynamic sensitivity tuning 30

e
Effort (search effort) 2, 5, 22, 26, 56
Entropy 9, 80
Epoch 65, 68, 76, 87, 88, 94, 95
Error
false positive 16, 18, 22, 25, 30, 39, 55, 56, 73, 92
false negative 16, 39, 55, 56
mean square (MSE) 65, 86
temporal difference learning 60, 66, 68, 84, 87
Estimated
location 9, 10
probabilities 18
reward 52, 53
rule 7
Expected
information gain 69, 70, 77, 79
location (position) 61, 85
path length 67
payoff 67
probability map 70, 79
reward 60, 69, 82, 85

f
Foraging behavior (animal mimicking) 5
Forward and backward recursive equations 2
Function
activation 59, 65, 81, 86
detection 16
detection probability 7, 22, 48, 50, 81
loss 65, 86
mean square error (MSE) 65, 86
probability mass 8
reward 44–46, 53
sigmoid 65, 86
value 44–46, 52

g
Global probability map 21, 23, 26, 27, 29, 32–35, 38, 47
Gradient 8, 53
Gridded domain 6, 21, 28, 31, 51, 52, 78

h
Heuristic algorithms/approaches/methods 5, 23, 55, 67, 69, 73

i
Independent opinion pool 24
Information
lack 17
learned 63, 84
Information gain
accumulated 37–41, 71
actual 34, 35
Information-theoretic metrics 32
Intermittence 63

j
Jensen–Shannon divergence 32

k
Koopman random search formula 2, 22, 78
Kullback–Leibler distance 32, 43, 58, 79

l
Lagrange multipliers 2
Layer
hidden 59, 60, 65, 81, 86
input 59, 60, 65, 86
output 59, 60, 65, 82, 86
Learning
brute-force 34–38, 42, 43
deep Q- 23, 43, 47, 50, 55, 58–60, 77, 78, 81
model-based 60, 62–64
model-free 60–63
reinforcement 43, 45–47, 50, 52
supervised 47–50
unsupervised 47, 49, 50
Local decision-making 8
Local probability map 21, 26

m

Manhattan metric 7, 10
Map
  agent 26, 39, 41
  global 26, 29, 30, 39–41
  sensor 26, 34
Markov
  decision process (MDP) 44, 52
  process 6, 7, 17, 61
  property 62
Markovian targets 21
Mean-squared errors 65, 86
Metric 7, 10, 32
Mobile search agents 21
Moving target 6, 9, 16, 26, 49–51, 55, 71–73, 77, 78, 92–95

n

Network training 60, 65, 68, 69
Neural network 48, 53, 55, 59–61, 63–69, 71, 77, 81, 84, 86, 88, 95
  prediction 60, 61, 65–68, 81, 83–85, 87, 88
  target 60–62, 65, 66, 84, 85, 87, 88

o

Observed area 7–9, 12–18, 51
Occupancy grid approach 6, 21
Ohsumi model 11
One-step look-ahead heuristics 77
Ornstein metric 32
Overlap degree/factor 12, 13

p

Partial data sharing 27, 29, 30
Path planning/path-planning algorithms 44, 46
Payoff, expected 67
Peer-to-peer communication / networks 5
Policy 7, 16, 23
  detection 36, 37, 39, 40, 41
  optimal 44, 45, 47
  search 26, 27
Pollock approach 2
Potential cells (visibility analysis) 11–15
Probability
  expected 50
  of target location 14, 22, 50
  transition 6, 7, 9, 44–46, 52, 61
Probability map 21–25
Probabilistic search scheme 5
Pursuit-evasion game 50–52

q

Q-max algorithm 55, 58, 59, 61–73, 81

r

Random search formula (Koopman) 2, 22, 78
Random walk (Brownian motion) 9, 72
Reactive search / reactive algorithms 6, 31
Reinforcement learning 43, 45–47, 50, 52
Rendezvous 52
Reward 43–46, 53, 59, 67, 84
  accumulated 43, 52, 53, 58–60, 69–72, 82, 85, 87, 89
  expected 44, 52, 53, 58–60, 69, 70, 71, 82, 85
  immediate 61, 66
  informational 58, 79, 87
Rokhlin metric 32

s

Sensing 22, 26, 48, 50, 57, 61, 65, 78
Sensitivity 22, 25, 27–30, 40, 41, 56, 64, 65, 73–75, 78, 80, 89–94
Sensor
  fusion 25
  probability map 21, 23, 24, 26, 32
Shadowed space 10, 11, 13–15
SPL algorithm 55, 58, 67–75
Stage
  acting 47, 48, 50, 86
  learning 47, 48, 59, 63, 65, 83–86
Static target 9, 28, 45, 49–51, 53, 71–74, 78, 81, 82, 89–92
Strategy 23, 50
Supervised learning 47–50

t

Threshold probability 27, 29, 30
Topography 10, 12, 14–16
Training 47, 49, 55, 60, 64, 65, 67–69, 72, 83, 84, 87, 88, 94, 95

data 64, 65, 67, 68, 87, 88, 94
Trajectory 5, 15, 18, 23, 27, 31, 37, 44, 49, 51, 55, 58, 61, 64, 65, 74
Transition probabilities matrix 6, 7, 9, 45, 52

u

Unsupervised learning 47, 49, 50

v

Value
iteration 44–46, 77, 81
Q- 50, 61, 63, 77
Visibility 10–13, 16
Voronoi viii, 3
partitioning 77
region 77–83, 85, 87

Printed and bound by CPI Group (UK) Ltd, Croydon, CR0 4YY
17/09/2026
14964270-0008